CRACKPOT

or

GENIUS

A

Complete Guide

to the

Uncommon Art

of

Inventing

FRANCIS D. REYNOLDS

CHICAGO
REVIEW
PRESS

Library of Congress Cataloging-in-Publication Data

Reynolds, Francis D.
  Crackpot or genius? : a complete guide to the uncommon art of
  inventing / Francis D. Reynolds
    p.   cm.
  Includes bibliographical references and index.
  ISBN 1-55652-193-6
  1. Inventions.   I. Title
T212.R49   1993
608—dc20   93-22803
                                                           CIP

First edition
Published by Chicago Review Press, Incorporated
814 North Franklin Street, Chicago, Illinois, 60610

ISBN 1-55652-193-6
Printed in the United States of America

5 4 3 2 1

This book is posthumously dedicated to Coe E. Wescott,
lifelong friend, fellow engineer, inventor, and
my associate on several long technical projects,
the person with the best mind I have ever personally known.

Contents

Introduction

History books are full of interesting stories of famous inventions, but seldom is it mentioned that most of the great inventors died poor.

In my teaching, consulting, and reading, as well as in connection with my own inventive efforts, I have become painfully aware of the overly optimistic and unrealistic views of inventing held by most laypeople and by beginner inventors.

There are several reasons for these distortions of reality. Inventing is a glamorous-sounding activity that makes news. The news accounts all too often emphasize the glamour but speak little of the difficulties.

Newspapers and magazines report on the successful inventions, but not on the failures. Failures aren't news, in most cases. This leaves the reading public with the impression that inventing is an easy way to make a fortune. We are not told that only one good idea in two or three thousand ever makes it to the marketplace. Nor do most people know that only 1 to 2 percent of all patents ever pay for themselves, and that only one patent in seven hundred makes big money.

Books on how to invent, how to get a patent, and how to sell your invention are for the most part designed to sell books. If a book tells its potential readers that they aren't apt to make it in the invention game, the book might not be popular. I would rather be honest. I'm telling it like it is.

The authors of other books on inventing are frequently patent attorneys or professional writers. Noninventors are not well qualified to tell others how to invent because they have never done it themselves. Patent attorneys may work closely with inventors, but not closely enough.

Lawyer-authors tend to cover their favorite subject, the patent system, in more detail than the average inventor or layperson is interested in. Lawyers are also inclined to overstate the need for legal services.

But the reverse is also occasionally true. I have a book written by a patent attorney on how to get your own patent without the help of a professional. Apparently in order to sell books, the author gives the impression that it is relatively easy to write your own patent application and negotiate it with the patent office. From all the authorities I have heard, and from my own experience, I know that getting a worthwhile patent without professional aid is extremely difficult and certainly not to be recommended.

So, this book is somewhat of an exposé. Where something stinks, I say so. I have tried to expose the companies that rip off unsuspecting inventors, and also expose the few crooked or ignorant inventors who rob careless buyers and investors.

For the most part, however, I have no axes to grind. The market for private inventions in this age is very poor, but that is no one's fault—it is simply a fact of the marketplace.

I am not concerned that the frankness of this book will discourage many inventors. The need to invent is born into some of us, and no book is going to make a true inventor give it up. The book will, however, acquaint new inventors with the problems of inventing and the odds of success, and hopefully save many of them both money and time.

At first this work was to be a handbook, text, or guide for beginning inventors. As the book has evolved, it is still for the inventor, but for others as well. It has grown far broader in scope and will appeal to a wide readership.

I like to invent, but I also find great satisfaction in teaching, lecturing, and acting as a consultant to other inventors. And the nature of the inventing game itself fascinates me. I have become an observer and student of the invention process, in addition to being one of the participants.

I have been both a corporate and a private inventor, but the corporate side of inventing is controlled by the rules of the various corporations. Therefore, this work will largely ignore that field. It will primarily address the private inventor and those with other interests in inventing.

One last thing before we serve the main course: I have covered a lot of different invention-related subjects in the book, including a little science, some language, perpetual-motion machines, advice to parents of inventive children, and so on. I wrote about those things because they interested me. I could do that because it is my book. While I was writing I was the boss.

But now you are the boss. If you start a chapter and don't like it, you can skip it and try for one you may like better. Different things for different people. Enjoy . . . and possibly even profit.

1

Can Inventing Be Taught?

When someone hears that I teach a course on inventing, they sometimes question the logic of it. "Inventors are born, not made—so how can you teach inventing?" This is a good question, and I have a good answer.

Yes, some of us have strong native curiosity, perseverance, creativity, a tendency to analyze and an urge to try things, dissatisfaction with hardware that doesn't work well, willingness to take risks, courage to challenge the experts, and desire to improve on the physical status quo. We also may be lazy and looking for an easier way.

These traits all lead to innovation. We were born to be inventors. Some other people have a strong sense of rhythm, the ability to carry a tune, and an ear for pitch. These people were born to be musicians. Still others are unusually well coordinated, have strong bodies, and a love for physical games. They were born to be athletes.

But did you ever hear of a concert pianist who didn't study music, have a teacher, and practice, practice, practice? Have there been any Olympic gold-medal winners who never had a coach and who didn't constantly work at their specialty?

Some people are born to be inventors, but before they can be effective at it they must *learn* to be inventors and to *practice* the art.

The old saying "Practice makes perfect" is of course exaggerated, since perfection is unachievable, but practice provides tremendous improvement in most fields. Inventing is no exception.

The ability of the human mind and body to continue to learn and improve amazes me. We can get better and better at most anything we set our efforts to. My first invention (a magneto) was in 1942, and I have been getting better at inventing ever since.

An engineer friend of mine won a major paper-airplane building and flying contest some years back. I asked him later how he did it, expecting to learn of some innovation of his in high-tech paper-airplane design. He admitted he had worked out a good design, but he said that practice in adjusting and flying his airplanes was the important part. My friend, a Japanese American, said that he was forced to spend years in a World War II internment camp as a young man. There was very little to do in the camp, and he got a lot of practice in building and flying paper airplanes.

Invention is sometimes called a science. It is really an art, but much invention requires extensive use of one or more of the sciences.

I am an engineer, and, as such, my personal inventions have been mostly technical or semitechnical in nature. On the following pages I therefore tend to lean toward hardware products and technical inventions, and many of the examples will be in those areas. However, the information, procedures, and lessons presented will apply to all types of inventions.

There is an old adage (which I just heard): A writer transmits, a reader receives. But unless the writer and reader are tuned to the same wavelength, all that one hears is static. The readers of this book are going to be tuned to several different wavelengths, however. I will try to stay tuned to all of you as much as possible.

I recognize that some readers of the book will have little formal education, while others will have advanced degrees. When an uncommon word seems to be the most appropriate choice, it will usually be defined for readers who may be unfamiliar with it.

The broad spread of education and experience in the readership requires that I compromise in the depth of coverage of the various topics. If you find some of it too elementary, please remember that I'm not talking down to you; I'm trying to help those with less knowledge understand the basics. Likewise, if you find a discussion of things you don't understand, be patient—we won't be above your head for long.

WOMEN INVENTORS

In the past, inventing was almost exclusively a man's game, like many other fields once strictly masculine. Several years ago I received a phone call from a woman named Autumn Stanley, who was trying to write a book on women inventors and was becoming increasingly frustrated in the effort because she was unable to find many women inventors to write about. I was able to give her some names new to her, but not many. I recently read that she finished the book, titled *Mothers and Daughters of Invention*, and it was published in late 1992 by Scarecrow Press.

In 1967, only one U.S. patent in a thousand was issued to a woman. That was far higher than the percentage a hundred years earlier, and far lower than the percentage of patents going to women today.

The students in my invention classes probably average 25 percent women. The clients in my invention consulting business also average around 25 percent women. About 15 percent of the people earning bachelor's degrees in engineering these days are women. The percentage of women inventors will continue to rise for many years to come.

MAKING A PROFIT

Students who acquire an engineering degree, or become lawyers or doctors, have no guarantees that they will be able to make a living in their field. In inventing, not only is there no guarantee of success, there is a high probability of failure in terms of financial gain. In this book I will try to teach you how to invent profitably, but I will also tell you what I know about other invention-related subjects.

Most experienced inventors invent because they love the challenge and because they can't help themselves. I think all inventors also hope for recognition and fame and want to be of service to others.

Probably most inventors initially think they will be able to make money, and many expect to be able to make a living as a full-time inventor, but very, very few ever achieve that goal. The reasons are many and complex and will be addressed in detail as we progress through the book.

I know of very few inventors nationally whose livelihood is earned by inventing, but I know none personally. Over the years I have met two inventors who claimed to be making their living by inventing, but as I got to know them it was evident, in each case, that their spouses were supporting their families with a different job.

A high percentage of inventors have patents, but as we shall discuss in later chapters, the great majority of patents don't even pay for themselves. According to Victor Di Meo, an invention consultant, less than 1 percent of the patents issued are ever commercially successful. Data published by the American Society of Inventors shows that only one in seventy patents breaks even.

Most of the famous inventors in history either died poor or, if they were financially comfortable at the end, they made their money in ways other than inventing. The few wealthy inventors whom many of us could name usually profited by manufacturing inventions rather than selling the rights to their inventions. In other words, they were successful businessmen in the marketplace as well as successful inventors in the history books.

For example, George Westinghouse is known as the inventor of the air brake for railroad trains. He received over a hundred patents, but made his money as an entrepreneur. The Westinghouse Electric and Manufacturing Company, which he founded, bought out the AC motor, alternator, and transformer patents of Nikola Tesla, a far more important inventor than Westinghouse, and paid very little for them. Tesla was in debt most of his life and died with nothing.

BACKGROUND FOR THE BOOK

I am a longtime inventor as well as an aerospace engineer, but I have discovered that I am especially fascinated by the inventive process.

It is said that the best way to learn is to teach. I have learned far more about the invention game than I would ever have known had I not taught the subject for sixteen years and been a consultant in the field.

Much of the material in this book is original, but where well-known concepts are discussed you will sometimes find different approaches and viewpoints that have evolved out of my teaching.

For the most part, writing the book was simply a matter of transforming the verbal material used in my classes and lectures into writing. The class presentations have been constantly updated over the years to better meet the needs of students and to include new material and thoughts as my knowledge of the subject expanded. As a result, this book, although a first edition, is comparable in clarity and completeness to a later edition.

A bibliography of fifty-five books on or related to invention is included. These books are in my personal library and present a good cross section of the field. I have added a brief evaluation of each book as an aid to choosing further reading. If you would like to read one of these books but find it out of print, try a bookstore that carries out-of-print books or a library. Most libraries keep books of this type a long time.

2

How to Rear Young Inventors

It is a well-known fact that the influence of parents on children is unpredictable, and the parents are far from the only influence, so the following advice is subject to the usual disclaimers. If you have an average family, not all of your children will turn out as you expect them to or hope they will.

There are several categories we might establish here: the parents who think their child was born to be an inventor but are wrong, the parents who think so and are right, the parents who want their inventive child to be something else, the parents who have an inventive child but don't realize it, and those parents who don't give a darn.

The last category is not amenable to any advice we might give here. This chapter will address the majority: parents who have some interest in their children.

LEAVE THEM ALONE

Certainly the young need nurturing, guiding, and educating, but the direction in which the guidance and education lead is critical—and frequently wrong for nurturing creativity.

If a parent is an inventor, scientist, or engineer, he or she will usually be delighted when a son or daughter shows similar traits, and there will be no problem. If mom is a musician and her daughter seems to be interested only in building things or reading science books, there may be a problem.

I enjoy working with bright youngsters to help them develop their creative, scientific, and engineering talents, but I would never try to make a technical person out of someone who has no inclination toward it.

If your children or anyone else's children are not interested in what you want to teach them, leave them alone in that respect. It is wrong to force people of any age into molds they do not fit. It may be a disappointment to a mother or father who is a physician to find that his or her child wants to be a plumber, but that's the way it goes sometimes.

Sure, young children are probably too inexperienced to know what they really want, let alone what is "good for them," so a lot of unbiased guidance is called for, but don't push. Much better a disappointed parent than a frustrated misfit offspring.

The answer to the question "What are you going to be when you grow up?" is apt to change several times as a child matures. Each new answer will be more realistic than the one before it. Even a young adult's final answer may not be the optimum one for him or her, but the choice should be his or hers.

When I was six I wanted to be "the man who puts tar on the streets." I had already felt a strong love for machines, and this street worker got to operate the biggest, noisiest, dirtiest, and therefore most glamorous machine I had seen, so my career goal made sense to me at that time.

When I learned about airplanes I wanted to be a pilot. My father was an electrician, so later I wanted to be that. I don't know how old I was when I found out what an engineer did, but from then on I was going to be an engineer, and I was right. My father was an inventor in his spare time. I feel the same urges.

How does a musician give unbiased guidance to an inventive son or daughter? It's tough today, and complete neutrality is probably impossible. Some bias is bound to be present because of the nature of human nature, but parents and teachers should try their best.

Of course, it is fortunate when the child seems to show the same interests as one or both parents, since the resulting home education will then be very valuable, and the child will probably receive a lot more attention, and even more love.

I would have hated it if my father had been interested in organized sports instead of making things. I was never on a Little League team, and never wanted to be. I did get plenty of exercise though. I was a gymnast, diver, and pole-vaulter—sports I chose for myself.

The experts are still debating over the relative roles of heredity and environment in determining what we turn out to be. Both play a significant role. Siblings more often than not have different interests and different natural abilities. That fact can only be explained by the laws of genetics, since siblings are reared in the same environment in most cases.

There is also no question that the home environment and education of a child are invaluable. I'm sure I was born with a creative mind and an inclination toward technical things, but I would be a quite different and lesser person if not for my father. He was inventive—interested in technology and in building things. He was also interested in showing me what he was doing and in working with me.

Education in the schools is equally important to a child. I regret to observe that education at the primary and high school levels has deteriorated since I was young, so home education is now even more important.

Unfortunately, home education has also deteriorated in a large percentage of homes. I could only speculate on all the reasons for this, but one of the most important seems to be that many parents spend less time with their children these days. Television, video games, and street gangs are not adequate substitutes for parental attention.

THE TOYS TO GIVE

The best gifts to give inventive children are tools to do whatever creative things they seem interested in. Don't buy "toy tools" that are too dull or otherwise unsuited for real work. After tools, give kids model kits or other raw materials that they can use to make things. The best "toys" to buy a young inventor are metal Erector sets, chemistry sets, or electronic experiment and construction sets, depending upon the interests of the child. Radio Shack stores carry electronic sets.

By the way, Erector sets are far better than other construction toys for the mechanically inclined child because the projects bolt together and they operate. The young inventor can learn about shafts, wheels, gears, pulleys, belts, motors, levers, trusses, etc. A close twin of the Erector set is the Mechano set. (Those who grew up with Mechano think it is better than Erector.)

Erector sets are not popular with impatient children because the projects take time to build, but impatience is not a sign of an inventive child. Lego sets, Lincoln Logs, and similar toys are intended for younger children. They build fast and develop architectural talents, but they are not mechanical.

Tinker Toy is on a middle ground. It builds faster than Erector and is somewhat mechanical (for younger children), but it doesn't teach nearly as much and doesn't permit the kinds of junior inventions that Erector does. Tinker Toy projects also tend to fall apart when operated, while Erector projects stay bolted together. Tinker Toy is a child's toy; Erector is a construction set for all ages.

I still have my sixty-year-old Erector set, and I even use it occasionally for building a first prototype of an invention or for evaluating a mechanical concept. (My set is not for sale; you will have to buy a new one.) A. C. Gilbert, the inventor of the Erector set and chemistry sets, has contributed much to the education of generations of young scientists, mechanics, engineers, and inventors. Thank you, Mr. Gilbert.

MUSIC LESSONS

If you think music lessons, ballet, or anything else would be good for your kid, by all means try it. Your decision may be right for your child, but be alert to the possibility that you are wrong.

My mother signed me up for piano lessons once. I appreciate the interest she had in my future, but I didn't like piano lessons. I didn't practice. I'm not very musical, and I had things to do that were much more important to me than playing the piano. I soon won and the piano lessons stopped.

Was I an ornery kid who didn't know what was good for him? I don't think so. I believe that although I was too young to put it into words, I instinctively knew that I wasn't cut out to be a musician, even a social one. My drives were elsewhere, and I was defending my right to shape my own future.

I doubt if Mom saw it that way when she let me off the hook, but her decision to stop the music lessons was right. She later admired me for my technical and inventive accomplishments and didn't seem to mind that I couldn't play the piano.

SUPPORT AN INVENTIVE CHILD

If I had been spending all my time as a child playing video games (fortunately they didn't exist then) or "hanging out," I would defend Mom's right to try to redirect my energies to more worthwhile channels. But I wasn't wasting my time. I was building all kinds of childish constructions, getting library books on science and inventing, and trying to blow up or burn down the house with chemical and electrical experiments.

In other words I was learning, getting experience, and preparing for my future. Not consciously, of course. I was playing, doing what interested me and came naturally, just as a young animal plays to develop its muscles and its hunting skills.

So, my message is that if a kid is doing things that are not self-destructive and that could be useful in some field, there is an excellent chance that the child is acting out his or her predetermined destiny. Don't try to change that course any more than you would force a naturally left-handed child to use his or her right hand. Don't ask, "Why can't you stop reading that book [experimenting, or whatever] and go out to play?" The child *is* playing . . . and you don't want him or her to play less usefully.

However, the exclusion of a child from contacts with his or her peers is apt to result in a socially maladjusted person. Sometimes a kid needs to play with the gang. In my experience it was a select gang, however. My childhood friends were always kids who had similar interests. We invented and built things together; we didn't just kill time.

I found I disliked kids who did nothing or who needed to get into mischief. I had to travel farther to find friends with interests similar to mine, because they were in short supply, but I found them.

This is nothing new, of course. People always tend to associate with their own kind. Artists congregate, as do bird-watchers, engineers, and hot-rodders. This type of selective gregariousness is not only more satisfying than associating with outsiders, but it is an important source of mutual education in one's chosen field of interest.

In high school and college the future inventor should take lots of math, science, mechanical drawing, and shop courses. For general inventing, engineering is the most useful college major, but inventors of medical instruments are usually doctors, inventors of garment inventions are usually clothing designers or garment workers, and so on.

But, Mom and Dad, this doesn't necessarily mean that you must see that your inventive young adult takes these classes in school. If the child is motivated in creative, inventive, and scientific directions, he or she will usually sign up for the courses needed without any help from you.

In many cases the opposite situation will be the problem: the student will know what is best for him or her, but the parent will be applying pressure in other directions. Monitor, and step in when you are *sure* you are right, but accept the fact that in most of these cases the student knows best, not Mom or Dad. Let go, but don't abandon.

ON ACCEPTING YOUR INVENTOR

If you find you have an inventive child, should you be happy or sorry for yourself? Should you feel proud or ashamed? As usual, "that depends." To help you sort out your feelings, here are a few thoughts on what your child's life may be like.

If you attach great importance to having lots of money and want your child to earn a lot, you may have a problem. Inventing is not apt to make a person rich, any more than acting, sports, prospecting, modeling, singing, or gambling is apt to make one rich. Invention is another game of high odds. A few inventors strike it rich, but very few.

But remember that inventing is very seldom a vocation—it usually remains an avocation or hobby. Since it is extremely difficult to make a living at just inventing, most inventors hold down a job to put bread on the table, and invent on the side.

Some jobs are more compatible with an inventing avocation than others. I am an engineer, which fits very well, especially since my inventive interests are mostly of an engineering or technical nature. In fact, the distinction between inventing and engineering is frequently narrow.

Engineers design things; inventors design *unique* things. Scientists discover facts of nature, and inventors and engineers use those facts.

Scientists discover and analyze natural things, and inventors discover and develop man-made things.

The building trades and general contracting are compatible with inventing. In fact, almost any field where hardware is involved will see its share of inventors, and their inventions will usually be improvements in their field of employment.

The educational level of inventors ranges from very low to postdoctoral, but the ones with the higher educations produce the most significant inventions. The intelligence of inventors will range from average to very high.

The social position of inventors can also be quite variable. Nikola Tesla was a graduate engineer and very popular in New York high society, and he dressed impeccably. Thomas Edison was crude, uneducated, and dressed sloppily. An inventor's interests and activities outside of inventing may be broad or narrow.

For the most part, inventors are happy and self-motivated people. They usually enjoy any publicity they receive in connection with their inventions, but they also may be frustrated by lack of acceptance of their ideas and by their inability to make money with their inventions.

In my observation, inventors are no more apt to be good or bad spouses and parents than other people are. Inventors come in all shades. Of course, an obsessive inventor may be so wrapped up in work that he or she neglects the spouse and children, but that problem is just as prevalent with obsessive doctors, lawyers, managers, or obsessive anything else.

What constitutes success in inventing? Making a lot of money on inventions is the most obvious answer, but because so very few inventors ever do, the number of successful inventors would be extremely limited by this criterion.

The number of patents earned is another measure of success, but again a poor one, because most inventions never make it to the marketplace, patented or not. Patents also vary widely in worth. One strong patent that fills a real need may be worth ten thousand ordinary patents.

To me, a successful inventor is one who contributes to the world, or simply improves one's own existence in inventive ways, whether or not he or she receives any recognition or compensation for those contributions. By that definition there are countless successful inventors.

3

The World Dislikes Inventors

We needn't spend time on the facts that great inventors go down in history, they are widely admired, books are written and movies are made about them, and creative small children worship them. After all, inventors have made life easier and fuller for us. They gave us such wonderful things as the automobile, airplane, telephone, radio, computer, refrigerator, you name it. Yes, inventors are loved. That is doubtless one of the attractions of inventing.

The reasons why inventors are also disliked part of the time by some people and some groups are a little less obvious, so let's look at the problems. As a simple start, inventors also gave us television (not all good), computer and video games (less good), guns, atom and hydrogen bombs, air pollution, nonbiodegradable plastics, liquor, cigarettes, illegal drugs, and gambling devices. We shouldn't blame the inventors for our misuse of these inventions, but it can be said that if it weren't for the inventors we wouldn't have the problems.

Many new inventions and ideas are counter to certain established thoughts and practices. New things may upset our habit patterns, rob us of traditions, compel us to learn new skills, make us different from the crowd, threaten our security, force us to think, or make something we cherish obsolete.

13

A sign in hotel rooms in 1892 read, "This room is equipped with Edison Electric Light. Do not attempt to light with a match. Simply turn key on wall by the door. The use of electricity for lighting is in no way harmful to health, nor does it affect the soundness of sleep." At that time that must have been about as confusing and upsetting as the instructions for programming a video recorder are today.

RELIGIONS SOMETIMES DISLIKE INVENTORS

In the past, particularly, scientists discovered many facts of nature and the universe that conflicted with accepted "facts" and sacred "truths" as propounded by religious leaders of all faiths. We are reminded of Galileo and his persecution by the Church for insisting that the planets revolve around the sun. There is still controversy between religion and science in such areas as evolution. Nor is the picture clear-cut; the scientists don't, and never will, have all the answers. Neither do any of the religions.

Certain religious groups, such as the Amish, prefer "the old ways" and reject modern inventions for the most part.

ORGANIZED LABOR DISLIKES INVENTORS

Inventions have made life much easier for the worker by providing machines to do most of the heavy work, giving him or her a comfortable working environment, and so on. But those facts tend to be forgotten by the people who have just lost their jobs because a new machine can do the job better and cheaper.

In the long haul, expanding technology puts more people to work, but the inevitable local and usually temporary loss of jobs due to technological progress cannot be ignored. We don't have a perfect economic system, and probably never will, and inventors sometimes get blamed for certain economic problems instead of the system.

ORGANIZATIONS DISLIKE INVENTORS

Work in any large organization can be frustrating for creative and inventive people. Theodore Levitt wrote in the *Harvard Business Review*:

> Organization and creativity do not go together, while organization and conformity do. The purpose of organization is to achieve the kind and degree of order and conformity necessary to do a particular job. The organization exists to restrict and channel the range of individual actions and behavior into a predictable and knowable routine. Without organization there would be chaos. Organization exists in order to create the amount and kind of inflexibility that are necessary to get the most pressing job done efficiently and on time. Organization must be inflexible to preserve order.
>
> Creativity and innovation disturb that order. Hence, organization tends to be inhospitable to creativity and innovation, though without them it would eventually perish.

That last concept may be surprising to those who hadn't thought about it, since most organizations claim to be strongly in favor of progress, and many promote it by such things as employee suggestion systems and invention awards. Organizations are interested in progress, but at the same time they subconsciously fear it because progress disturbs order.

Management consultant Peter Drucker said, "The hardest organizations to change are the ones that are large, successful, and old." Large corporations are much more set in their ways than small ones. Large and well-established governments are also very difficult to change. On the whole that is fortunate, because not all change is good. Proposed changes should be required to pass many tests.

Understanding why organizations resist change will not make the problem for the inventive employee go away, but it will help. The inventor who can't get his or her boss interested in a great new invention—which the inventor is sure will improve productivity, add a profitable new product, or do other great things—will always have a problem. If the inventor understands the organizational pressures on the boss, however, he or she is better able to try again to present the invention, or perhaps to see the boss's side of the picture better or even conclude that adopting the invention would not be advisable for the organization.

New ideas take time and money to develop. New ideas meet opposition from many quarters. The risk of failure is great with new ideas. If an existing method, design, process, or procedure is reasonably satisfactory, use it. Don't improve something until it no longer works. Don't reinvent the wheel.

It is easy to generate ideas. The hard parts are to reduce good ideas into practical hardware, to get it to the marketplace, and to sell it. You will see many speakers and writers on the subject of creativity who imply that the idea is the all-important thing, that if we have creativity, all of the financing, hard research and development work, manufacturing start-up problems, and marketing will take care of itself.

Not only are speakers and writers often guilty of such shallow positions, so are many would-be inventors. People with such attitudes have had little or no exposure to the real world of manufacturing and business. The statement "Ideas are a dime a dozen" is not popular with idea-generating people, but it carries a great deal of truth.

COMPANIES DISLIKE OUTSIDE INVENTORS

Most large companies receive a lot of mail from outside inventors who are trying to sell their inventions. Usually the companies are not interested. Usually companies wish these outsiders would go away, because they are an expensive nuisance. There are several reasons for this.

Most large companies have their own research and development organizations, and the creative people in these departments are usually far more qualified than outsiders to come up with viable new products or improvements for the company.

Not only are the insiders usually more qualified, but they have a vested interest. Why should the insiders accept outside inventions when they can invent whatever is needed better, and at the same time justify their own paychecks. The phenomenon is called "NIH," which stands for "not invented here."

As an example of the resulting odds, one study showed that the General Motors group that reviews outside suggestions looks at about sixty-five hundred per year and, on the average, accepts only *two* per year!

Another reason why companies tend to dislike outside ideas involves liability. In most cases companies will not agree to accept a confidential disclosure of an outside invention, and usually they won't listen at all, unless the inventor signs away all his or her rights to confidentiality or has applied for a patent.

An outside inventor sometimes may sue a company, claiming that it stole his or her invention. This is occasionally true, but more often the inexperienced inventor doesn't understand the system and is in error. But because the suits cost a company's time and money and can hurt its public image, it leans over backward to avoid the possibility.

In a frequent scenario, the inventor tells the company about the invention and the company rejects it, or the inventor thinks the company somehow learned about it. Years later the company puts a product on the market that looks to the inventor like his or her invention, so he or she complains to everyone who will listen and/or sues the company.

Several points need to be considered here. The company may well have already been working on the same or related ideas at the time of the disclosure, but it is not apt to tell an outside inventor what its future business plans are.

Also, small-time inventors are usually not experienced enough with the patent system to realize that the protection provided by most patents is very narrow. To the outside inventor, a company's new product may look broadly like the inventor's own invention, but a professional examination of the outside patent will usually show that what the company developed does not infringe on the outside patent.

Of course, if the "inventor" never got a patent on the idea, he or she has no legal rights and the company is free to use it with no obligation, even if the company first heard of the idea from the inventor. Ideas do not become "intellectual property" unless and until the government grants a patent, a trademark, a copyright, or recognizes a trade secret. (More about that in later chapters.)

The control system invented to operate this radio-controlled model boat later controlled a Boeing guided missile.

Of course, inventions are sometimes stolen from small inventors, but it is rare. Nevertheless, the media help perpetuate the myth that big companies frequently steal inventions from the little guy. Organizations are thereby unjustly accused and would-be inventors are misled.

Most major companies have a small brochure that they send to outside inventors who approach them. These brochures inform inventors of the companies' strict rules for considering outside inventions. As an example, a brochure from the Parker Pen Company reads in part, "Parker will be the sole judge as to the compensation, if any, paid to the submitter for an idea that is not covered by a valid United States patent."

There is a personal story I would like to tell you, concerning an outside invention and a big company, which illustrates the reluctance to seriously consider outside ideas.

I once became involved in a radio-controlled model boat project of the Seattle Model Yacht Club, and, about the same time, I was assigned as an engineer to the Boeing BOMARC guided-missile project. The first project was an off-hours hobby and the second was part of my professional career, but it wasn't long before the two activities merged.

Another Boeing engineer, Leroy Perkins, and I invented a twenty-channel control system specifically for the model boat. He and I submitted our invention to the Boeing patent staff under the terms of the company invention agreement that we had both signed. (More on invention agreements in Chapter 27.) The company kept it for a few days and sent it back to us with a note that Boeing had no interest in toy-boat control systems, therefore it was all ours.

Ten years later we took the model boat to England and won a world's championship with it.

In the meantime, back at Boeing, we were having trouble coming up with a good system for the midcourse guidance (radio control) of the BOMARC guided missile. It became evident to me that what "Perk" and I had invented to control the model boat could also control a guided missile and, in fact, was just what Boeing needed. I made a presentation to my management. They agreed and directed the Boeing patent staff to reacquire the rights to the model-boat control invention.

Perk and I signed a contract with Boeing, selling them the invention that they originally owned under the invention agreement, but gave away because the patent engineer didn't appreciate its significance. I became the manager in charge of the engineering development of the Boeing version of the device. Before the development was finished we were awarded three more patents on improvements to the invention.

According to a study, the use of that "toy boat" invention in the guided missile saved 350 vacuum tubes per missile. (This was before the availability of transistors and integrated circuits.)

Elsewhere in this book I mention that courage is a desirable trait for an inventor to have. This little story contains an example. When I was trying to convince my supervisor that Boeing needed our invention, he reminded me that the guided-missile control system had to work five times as fast as the system in the model boat, and he asked if our invention could run that fast. I had worried about that point already but tried not to show it. I said, "Sure boss, no problem!"

I lucked out and wasn't fired. In a month or two we had developed a high-speed version and successfully tested it at over twice the required speed.

The Boeing version of the toy-boat control invention

Note that the "outside invention" in this case had a much better chance than most would have had because, while it is true that the invention was made outside of the company and for a different purpose, the inventors were employed by and well known to the company. More important, this outside invention happened to fill a real and current need of the company. That will seldom be the case.

SPORTING GROUPS
DISLIKE INVENTORS

The potential for conflict between inventors and organizations isn't limited to commercial organizations. Most sporting organizations have rules to govern their sport. Invariably these rules are designed to limit or prevent progress in the equipment used by the sport.

For example, when aluminum pole-vaulting poles were proposed to replace bamboo poles, they were initially not allowed. When fiberglass poles were developed to replace aluminum, they were

The author's high-tech, radio-controlled sailboat was developed in a wind tunnel.

outlawed at first. When epoxy/carbon-fiber poles first came out, they were forbidden in competition. The reason for these restrictions was, of course, to prevent unfair advantages in the sport. Every competitor must have the new and better equipment before any can be allowed to use it.

Those of you who have followed the America's Cup sailboat races will recall that there have been many court battles over whether certain improvements in specific boats were legal under the existing rules.

I was active in radio-controlled model yacht racing in the 1950s. In 1952 I won the national championship in that sport, and I lost it the next year. That defeat launched me into an effort to design and develop a faster type of sailboat. The resulting boat, with a wing-type sail that I developed through extensive wind-tunnel testing, was better than conventional sailboats, but the Model Yacht Racing Association of America wouldn't let me race it in competition unless I accepted a severe handicap.

I am pleased to note that the America's Cup winner *Stars and Stripes* had a wing sail very similar to mine, but my radio-controlled model was sailing more than thirty-five years before the *Stars and Stripes*.

The characteristic of sporting organizations to discourage progress is, in my opinion, one reason why some inventors tend to be loners. If the group says, "You must play by our outmoded rules," the inventor loses interest in their game.

TEACHERS DISLIKE INVENTORS

This section is closely related to the earlier one on organizations, since teachers are a part of educational organizations. Teachers must follow a curriculum and cover a certain amount of material over a specified time period, so there cannot be unlimited flexibility. The student who creates a different answer than the one in the textbook will be discouraged by many teachers, even though the student's answer is right. It is a matter of learning facts versus learning to think. Both are important. In earlier times, the teaching of facts was paramount. Modern education is better balanced, but creative students are still often silenced instead of encouraged to express and examine their ideas.

It is natural for an instructor to expect all students to learn something in the way and at the time that he or she chooses to teach it. Unfortunately, creative minds don't always work that way. Thoughts come into the creative mind on a random basis; they can't be readily turned off and on to a formal schedule. The person who seldom gets an original thought but who shows cooperation and a good memory is frequently considered a better student than one who is just as bright and has much more creativity and individuality.

As the schoolteacher said, "Orville! Wilbur! Stop sailing those paper airplanes this instant!"

The teachers I remember and admired in grade school, high school, and the university were the ones who taught me ideas as well as facts. These included my parents, a grade school math teacher, a high school chemistry teacher, and several engineering professors. You good teachers know who you are. Thank you for encouraging us to think!

EXPERTS DISLIKE INVENTORS

Experts are people who know what can't be done. Inventors think things can be done. Therefore, inventors and experts will often disagree.

Let me tell you a true story, in which I was both the expert and the villain. In 1963, I was the manager of a Boeing electromechanical design group. One of our projects was to invent, design, and develop a means for getting strain-gage signals out from a gas turbine rotor spinning at 50,000 rpm.

We tried conventional slip rings but they weren't satisfactory, because at that speed they overheated and wore out very rapidly, and the electrical noise from the slip rings masked the small strain-gage signals. We were looking at more sophisticated ways of doing the job, such as optical or radio data transmission, but these would have been expensive to develop and had other problems.

At this time, Frank Maytone, one of my more creative engineers, came into my office and said, "Francis, I know how we can solve the problem. We can use slip rings, but we will fill the slip-ring case with water." I reminded him that these were electrical slip rings and that electricity and water don't get along well together. "The water would short out the circuits, and it wouldn't work," I informed Frank. He left my office quietly without arguing.

About a week later he came back and said, "It works." I asked him, "What works?" since I had forgotten about his worthless suggestion. Frank took me into the laboratory, where he had set up a test with water-flooded slip rings that he had somehow gotten built without my authorization.

They certainly did work—they transmitted the strain-gage signals perfectly at not only 50,000 rpm, but at 135,000 rpm! The presence of the water cooled the rings and brushes so they wouldn't overheat, it lubricated them so they wouldn't wear out, it cleaned them so the electrical contacts wouldn't become intermittent, and it damped their vibration so the brushes wouldn't bounce. It was a beautiful solution to the problem!

But what about my "expert" opinion that it wouldn't work? I had said the water would short out the circuits. Everybody knows that water will conduct electricity, which is why we shouldn't touch electrical appliances when we are in the bathtub.

Yes, water conducts electricity; salt water conducts it quite well. Fresh water is a fair insulator, however, and distilled water is a very good insulator. If the signals we had to measure had been high-voltage on high-resistance circuits, I would have been right. But these were low-voltage and low-resistance circuits. The shorting effect of fresh water on them was negligible. I had been too hasty in pronouncing judgment on Frank's radical idea.

We used Frank Maytone's flooded slip-rings invention very successfully at Boeing. Frank got a patent on it, which was licensed to an instrumentation company, and Boeing and Maytone collected royalties on its use in other applications. All this took place because Frank Maytone thought he was right and had the courage to work on the invention in spite of his supervisor's rejection of his idea.

4

Who Invented It?

THE TRANSISTOR

In school we were taught the names of the great inventors, or at least I was in the 1930s. We learned about Thomas Alva Edison, James Watt, Robert Fulton, Samuel B. Morse, the Wright brothers, Alexander Graham Bell, Eli Whitney, and many others. I admit I haven't read any elementary school textbooks lately, but I wonder if the modern ones tell our latest generation of students who the inventors of the transistor were.

The transistor is the basis for almost all modern electronics. Integrated circuits (silicon chips) use microtransistors by the hundreds. Without transistors we would have no modern radios, radars, portable telephones, televisions, computers, calculators, digital watches, satellites, automobiles, airplanes, fax machines, or even (perish the thought) video games. But who invented the transistor— the very heart of modern electronics?

The transistor was invented in the Bell Telephone Laboratories in 1948 by John Bardeen, Walter Brattain, and William Shockley. Thank you, gentlemen.

My daughter, Barbara, had Walter Brattain as a physics professor at Whitman College in the mid-1960s. She didn't find out until after

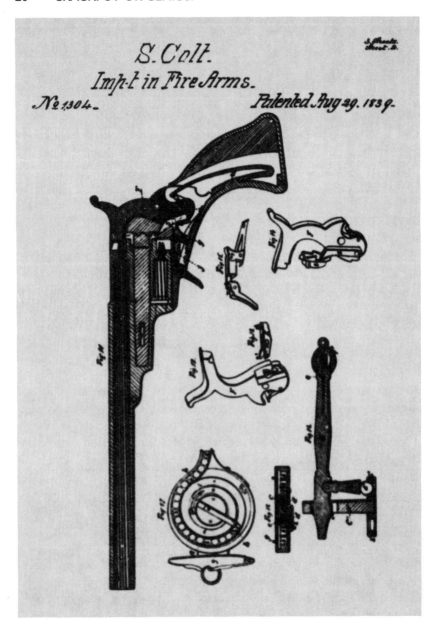

Colt patent

274. SOUND RECORDING & REPRODUCING

T A: EDISON.
Phonograph or Speaking Machine

No. 200,521.   Patented Feb. 19, 1878.

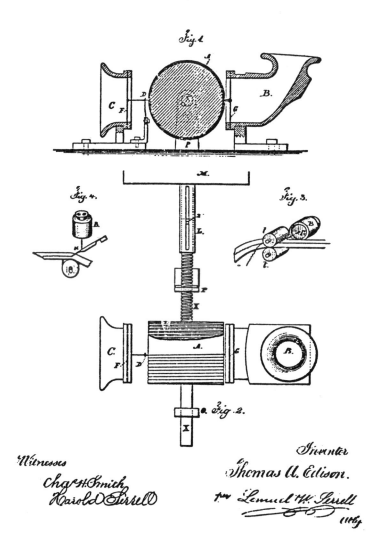

Inventor
Thomas A. Edison.
per Lemuel W. Serrell

Witnesses
Chas H. Smith
Harold Serrell

Edison phonograph patent

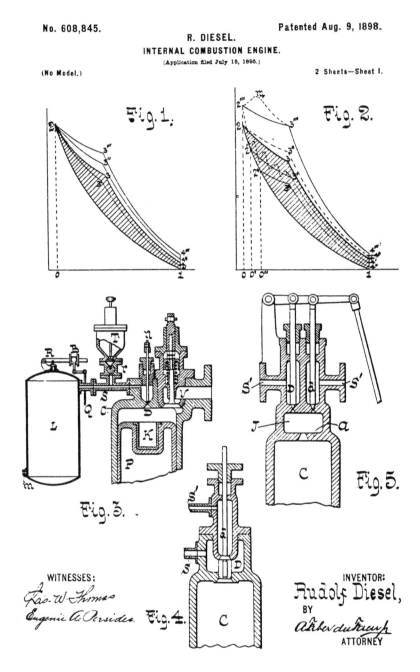

Diesel patent

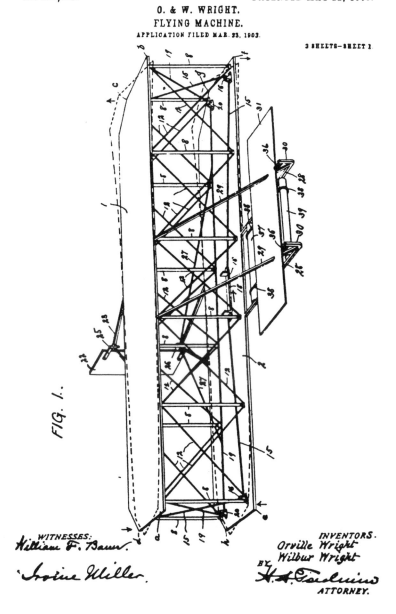

Wright Brothers patent

the course was over that her professor was a coinventor of perhaps the greatest invention of all time. Fame is fleeting.

THE AUTOMOBILE

Who invented the automobile? Henry Ford? Carl Benz? Gottlieb Daimler? Albert De Dion? Rene Panhard? Emile Levassor? Ransom Olds? Most of these entrepreneurs were in the automobile manufacturing business before 1900, and they all contributed to the development of the automobile, but they didn't invent it.

Joseph Cugnot built and demonstrated a steam-powered car in France about 1765. It would carry four people at two miles per hour. Siegfried Marcus made the first successful European car with an internal-combustion engine in 1875 in Vienna. Also about 1875, George Brayton ran an internal-combustion automobile in Philadelphia.

The first U.S. patent on an automobile was applied for in 1879 by George Selden, a lawyer who never built an automobile. He filed improvement revisions on his application for a period of years, and the patent was finally issued in 1895: U.S. patent number 549,160.

In his original patent Selden claimed, "The combination in a road vehicle equipped with an appropriate transmission, driving wheels, and steering, of an internal combustion engine of one or more cylinders, a fuel tank, a transmission shaft designed to run at a speed higher than that of the driving wheels, a clutch, and coachwork adapted for the transport of persons or goods." Cars today still fall under that wonderfully broad claim.

Selden collected royalties from various automobile manufacturers in the early 1900s, but a group of companies, including Ford, finally beat the Selden patent in the courts.

So who invented the automobile? Take your pick.

THE RADIO

History books give Guglielmo Marconi credit for inventing the "wireless," the original term for what we now call radio transmission; but the original wireless was Morse code only, no voice or music. And Marconi wasn't the first.

Michael Faraday discovered magnetic lines of force, which extend through the air. James Clerk Maxwell developed the theory of

Henry Ford (left) and Thomas Edison (right). *Photo courtesy of the U.S. Department of the Interior, National Park Service, Edison National Historic Site.*

electromagnetic radiation. In 1888, Heinrich Hertz produced radio waves in the laboratory and received them from across the room.

Marconi received a message from nine miles away in 1896. A few years later he received the single letter *S* across the Atlantic Ocean. The inventor of alternating-current power systems, Nikola Tesla, claimed to be the father of radio and competed with Marconi, but very unsuccessfully.

Thomas Edison discovered the "Edison effect" (thermionic emission current in a vacuum) but didn't understand it. John Fleming in England used the Edison effect to invent the thermionic "valve," or diode vacuum tube, in 1904.

Lee De Forest, back in the United States, added a "grid" to the Fleming diode to invent the triode vacuum tube, the first tube capable of amplifying a signal. But De Forest understood his invention only poorly and wasn't able to use it in radios very effectively.

The brilliant electrical engineer Edwin Howard Armstrong was the first to really put the vacuum tube to work. Armstrong invented

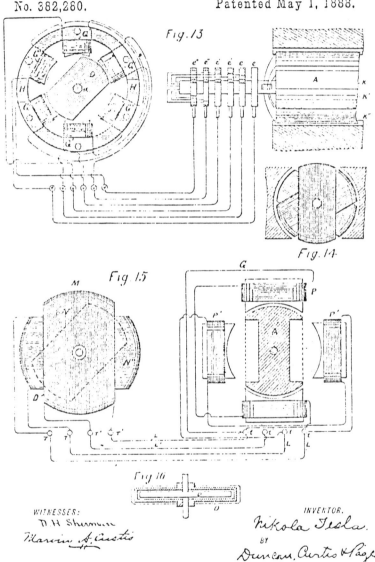

Tesla patent

most of the basic circuits of radio, including positive feedback (regeneration), which he patented in 1913; oscillation with a vacuum-tube circuit; the superregenerative circuit, which greatly increased range; the superheterodyne circuit, which is still the basic circuit of all radio receivers; and FM radio.

Neither De Forest nor Armstrong was the first to transmit voice and music. That achievement was made by Reginald Fessenden, another electrical engineer, in 1906. Earlier, Fessenden worked for Westinghouse, Edison, and General Electric and was an engineering professor at Purdue University.

David Sarnoff did some personal tinkering in the early days of radio, but he is most noted as an entrepreneur and business leader in the field of radio, and later television. Armstrong worked for Sarnoff for many years when Sarnoff was head of RCA (which absorbed the Marconi company). Sarnoff later formed the National Broadcasting Corporation (NBC).

All of these men were ambitious, egotistical, and jealous. There were many long and bitter court battles between De Forest and Armstrong, and later between Armstrong and Sarnoff.

So who invented radio? All of the above, plus the more recent inventors of the transistor. I pick Edwin Howard Armstrong as the most important inventor of radio. He contributed the most in the way of science and engineering, and most of his circuits are still used. Marconi's code wireless and De Forest's vacuum tube are dead.

It is interesting to observe that galena crystals were used in early "crystal sets." The vacuum tube put the crystal out of business. But semiconductors have put the vacuum tube out of business, and they use modern synthetic crystals (doped silicon instead of galena). We have come full circle.

5

Amazing Toy Inventions

I love toys, especially creative technical toys, and double-especially toy inventions that shouldn't work but do. Here are several fascinating ones for your mental stimulation.

THE PERPETUAL TOP

In gift catalogs or novelty stores you may have seen a little top that balances on a plastic base and spins and spins and spins, apparently forever. There are other versions of the same concept: a pendulum appears to swing forever without any driving force, a wheel seems to roll on a ramp forever, and so on.

To those of us who love gadgets and know that perpetual motion is impossible, these toys are engrossing. How do they work? They are really quite simple. Most mysteries seem simple when we know the answer.

As you may have guessed, they require a battery for motive power, and therefore they are not violating any laws of physics. But where is the battery, and how does it make the device work? The battery is in the base, connected to an electromagnet. There is also a permanent magnet in the rotating or moving part of the toy.

34

The toy top that spins for days is a brushless DC electric motor.

Direct-current (DC) motors are made with electromagnets and permanent magnets and are powered by an energy source such as a battery. But those of you who are familiar with such things will tell me that a DC motor has wires and requires brushes and a commutator. Those essential parts seem to be missing in these toys. The forever-spinning top *is* a special DC motor, but brushes and a regular commutator are not required because the commutation is accomplished electronically.

There is a transistor in series with the battery and the electromagnet in the base. As a pole of the permanent magnet in the top swings past the upper pole of the electromagnet below it, it induces a voltage in the coil of the electromagnet. This voltage turns on the transistor, which in turn feeds battery power to the coil to energize the electromagnet. The resultant magnetic force interacts with the field of the permanent magnet to keep the top spinning. It will run for several days until the battery is exhausted. Now you know.

PORTABLE SLOPE SOARING

One of my hobbies is the design, construction, and flying of radio-controlled model airplanes. One form of that hobby consists of flying radio-controlled gliders in front of a cliff or slope, where wind approaching the slope is forced upward by the slope. Under good conditions it is possible to keep a glider aloft in such updrafts for hours on end, to do aerobatics, and to race against other RC gliders flying on the same slope.

The clever toy I am about to describe is also a remote-controlled slope-soaring glider that can stay up for a long time, but it is very inexpensive, very small, very light, has no radio, and can be flown indoors.

The secret is disclosed in the title of this section: the "slope" is *portable*. Instead of a hill or cliff, our slope consists of a sheet of cardboard or a writing tablet. It is held at an angle in front of the operator or "pilot," and the little glider is launched above and in front of the cardboard sheet. The pilot now walks forward, which causes the air around the glider to deflect upward over the edge of the cardboard, providing lift. The glider may be turned at the will of the pilot, by properly maneuvering the sheet.

I have seen one of these very interesting little gliders flown. They were invented in 1977 by Parker MacCready, age 17, and Tyler MacCready, age 15, sons of inventor Paul MacCready. Invention does often run in families. My father was an inventor, I am, and so is my son.

THE WINDMILL SAILING CART

Good ice boats and wheeled sailing carts can go much faster than the wind in a reach. (That is sailor-talk for sailing a crosswind course.) They can't sail straight into the wind at all, however, and they can't sail straight downwind faster than the wind. At least the laws of nature say they can't, and I'm not about to argue with Mother Nature. Even people who think perpetual motion should be possible can't see how such a sailing vehicle could accomplish these two things.

A few years ago a group of engineers were wondering about these apparent sailing limitations. Dr. Andy Bauer, an aerodynamicist, at that time with Douglas aircraft, was convinced that machines could be built that could sail directly into the wind and sail directly downwind faster than the wind. He got so much flak from his technical colleagues that he designed and built a "windmill sailing cart" to prove his point. It did what he claimed, and it is *not* perpetual

Andy Bauer with his Windmill Sailing Cart, which sails downwind faster than the wind! *Photo courtesy of Andy Bauer.*

motion! The laws of thermodynamics and conservation of energy are still intact.

In addition to proving his point by test, Dr. Bauer wrote a series of technical papers on the concept, proving it mathematically and analytically. These papers were presented to a specialty group of the American Institute of Aeronautics and Astronautics.

Many of you will remember the name of Paul MacCready, whom I mentioned earlier. He is the engineer/scientist who was behind the *Gossamer Condor* and *Gossamer Albatross*, the human-powered flight machines that won one Kremer Prize for flying a figure-eight course and a second Kremer Prize for flying across the English Channel. Both used a power plant called Bryan Allen, which developed only one personpower.

Anyway, Dr. MacCready heard about and admired Andy Bauer's unique sailing-machine concept and mentioned it in some of his public lectures. That is how I heard of it. I have since discussed the concept with both Andy Bauer and Paul MacCready.

Let's talk about how the windmill sailing cart works. We have a four-wheeled vehicle that has one or more of the wheels coupled to a transmission system, as in automotive practice. The transmission is in turn coupled to the power plant, as you might expect. The power plant is a bit unorthodox, however. It is a windmill, a mobile windmill.

In going directly into the wind, which most people will accept as possible with this machine, the wind drives the windmill, which powers the wheels, which moves the cart. We obviously have to develop more traction force at the drive wheels than we have wind force against the windmill. That can be achieved by using gearing to give us a mechanical advantage. Gears can increase torque, and therefore force, at the expense of rpm (or vice versa).

However, the claim that throws most people is that this type of machine can travel faster than the wind downwind, driven by that selfsame wind. It especially throws those who don't believe in witchcraft. If you are currently convinced that this claim is impossible, you too have been thrown. But some wise man said, "When one way doesn't look right, look for another way."

Another way: Instead of the windmill driving the wheels in the downwind mode, let's see if the wheels can drive the windmill. The wheels are spinning because the wind is pushing the machine downwind, so let's use some power from the wheels to turn the windmill. But here we had better change the name of our aerodynamic screw device from "windmill" to "propeller," since we are going to use it to develop thrust instead of develop mechanical power. This is the most essential point to remember in understanding the concept. I repeat: In sailing directly downwind faster than the wind, the wheels must drive the propeller!

To further paint a mental picture for you, recognize that if the sailing cart had no wheel friction, it would drift downwind at exactly wind speed, like a drifting balloon, and it would experience nothing but still air. But by turning the propeller slowly in that still air we generate a little bit of thrust, and that thrust will move the vehicle with respect to the still air. In other words, it will go faster than the wind.

But, you vehemently object, if we load the wheels, that "braking" force will slow up the cart. Sure, I freely admit, it will *tend* to slow the vehicle, but this decelerating force will be less than the accelerating thrust we are going to get from the propeller, again because of the mechanical advantage in the transmission system. Gears are great.

This is a surprisingly difficult and controversial little technical problem. One needs to have an open mind before a technical proof can be developed. If one is convinced that it is impossible to extract useful energy from the wind when we are traveling faster than the wind, one tends to devote all his or her energy to defending the "it-can't-work" position, instead of trying to understand how it does work.

An ice boat can sail faster than the wind at an angle to the wind, but our windmill sailing car can eliminate the angle of the course because the angle is built into the propeller blades. The blades follow a helical path with respect to the wind, which serves the same purpose as the angular path of the ice boat.

That is one way of understanding the problem, but each person I know of who has gotten the correct answers arrived at them in a somewhat different way. I predict that some of you will lose sleep tonight thinking about it. Those of you who eventually figure it out may not do so until after days or weeks of studying. It is the toughest simple-looking technical puzzle I have seen.

THE POWERED POGO STICK

I guess we could say that all pogo sticks are powered during use—human powered. But this one is gasoline powered. I don't know who invented it, but the patent number is 2,929,459, if you want to look it up for yourself. It was made by the Chance Manufacturing Company of Wichita, Kansas, under the trademark "Hop Rod." Chance is one of the largest producers of carnival and amusement-park rides.

My unit is still in hopping condition, but Chance hasn't made them for about twenty years. I suspect they were withdrawn from the market because they are a bit dangerous to use. Mine threw me and gave me a sore elbow for awhile, and my son got a broken foot from it. No company could live with that potential in this age of rampant liability suits.

While the windmill sailing cart was very difficult to understand, the Hop Rod is easy. (Easy to understand, and fairly easy to ride.) The engine runs at about sixty hpm (hops per minute), and it gets very good mileage.

It lifts passive adults three or four inches on each hop, but light kids get much more altitude. If any rider jumps with the machine, he or she can go a lot higher. By locking the knees and letting the machine do all the work, the rider only needs to steer, by leaning in the direction of desired travel.

The gasoline engine itself is obviously a bit unusual. It has no crankshaft or other rotating parts. It is a two-stroke-cycle type and uses oil in the gasoline.

Ignition is by a primitive system that was used on some regular gas engines around 1900. The tubular post has eight C-size dry cells slipped into it. These are connected to the coil and to the spark plug. There are no breaker points in the usual sense. The coil has only a single winding; it is not a transformer as modern ignition coils are. The spark plug has only the center electrode, and it consists of a long spring wire that extends into the cylinder.

To operate this unorthodox collection of electrical parts, the rider jumps on the unit and the piston compresses the mixture. At the top of the stroke the piston makes electrical contact with the spark plug electrode, and the battery energizes the single-winding coil. As the piston starts to go down again (actually the cylinder goes up instead) the electrical connection at the spark plug electrode is broken. The magnetic flux in the coil core rapidly collapses and induces a high voltage in the winding, which causes a spark across the widening gap, which ignites the mixture: bang, bang, bang.

It is a lot more educational and better exercise than video games.

The lazy man's pogo stick

6

Crackpot Inventors

The invention game—hobby, avocation, business, obsession, or whatever it is—has its share of crackpots. These individuals add color to inventing and are generally harmless.

It is impossible to distinguish some future great inventors from the crackpots, however. Some of our most important inventions seemed so ridiculous at the time of their invention that the common person and also some of the "experts" labeled them impossible—flying and sending messages without wires, for instance.

The inventors of these things were originally labeled crackpots. The final criterion for judgment sometimes is, if it works the inventor is a genius, if it doesn't work the inventor is a crackpot. This is about as logical as saying if the patient lives the doctor is a genius, and if the patient dies the doctor is a quack.

True crackpots tend to give inventors in general a bad name, but we are big enough to weather that. However, some misguided would-be inventors cause harm to themselves and others. I say "would-be" inventors because most of their ideas don't work, and if something doesn't work it isn't an invention.

Many such inventors have been killed or injured in testing flying machines that don't—at least not for long. However, a far more common form of damage caused by incompetent inventors is financial.

Crackpot inventors are typically so obsessed that they will spend all of their own money on their ideas and also all of the money they can talk other people out of.

Such zealots are frequently very convincing; their enthusiasm is contagious. There are plenty of ignorant and average people who will fall for an apparent opportunity to "get rich" and promote a "wonderful new invention" at the same time.

Our laws try to protect the public from all charlatans who knowingly attempt to defraud, and there are charlatan "inventors." The regulations controlling stock promotions are strict. More common than the crook, however, is the ignorant inventor who truly believes in an unworkable or impractical idea.

PERPETUAL MOTION

As we will see in Chapter 7, perpetual motion is impossible, but some would-be inventors and some of the general public will never be convinced of that fact, so these pie-in-the-sky ideas will continue to be promoted.

"Perpetual motion" literally means moving forever, but that isn't what the term commonly means. The earth will move around the sun for a very long time, but it doesn't generate any energy in doing so. "Perpetual-motion machine" inventors are trying to get usable energy for free. They were more common a hundred years ago when science was less advanced, but we still have them with us.

Those of us who remember some physics or have studied engineering know that such machines can't work because they would violate the natural law that energy can be neither created nor destroyed. Energy can be converted to other forms of energy. Energy can also be converted into mass, or vice versa, in nuclear reactions. That is what Einstein's $E = MC^2$ equation is all about.

Even though perpetual-motion machines can't work, I am still interested in studying each new one that comes along, to see in detail why it won't work. To me, details are more challenging than generalities. In some clever PM-machine ideas it is a bit difficult to identify which step or steps in the proposed operation are false.

At an inventor's fair I participated in some years ago, I met a pseudoinventor who claimed he had a perpetual-motion machine. He even had models of it there. Of course, the models didn't work, but he assured me that was only because he hadn't gotten them adjusted yet.

His machine was a variation on a popular class of perpetual-motion schemes, in which one side of a vertical endless belt with floats on it is in a tank full of water, and the other side of the belt is in the air. The theory is that the side in the air has weight, but the side in the water has buoyancy, so the differential force is supposed to move the belt and turn its pulleys.

Analysis always shows it can't work, regardless of the design details, because in order for one side of the belt to enter the tank through some kind of a hydraulic seal at the bottom, it theoretically and actually must lose its buoyancy and weigh the same as the other side. The water must be able to get *under* an object before it has buoyancy. A loose cork placed at the spout of a funnel full of water is pushed down—it does not float up.

I tried to explain to the man at the inventor's fair why his machine couldn't work, but he would hear none of it. He had his own expert who said it *would* work. He gave me a copy of a paper from a "professional engineer" he had hired, in which the professional attested to the workability of the concept.

As a professional engineer myself, I am concerned with ethical conduct in the profession, so I looked up the PE and challenged him. He too would not be convinced, but I did warn him concerning his malpractice. I couldn't find his name in the roster of professional engineers for the state of Washington, so perhaps he was guilty of impersonation as well as ignorance.

The sketch on the next page shows a perpetual-motion scheme that has as much chance of working as any other. You will note that on the left-hand side of the wheel there are six-pound weights, but by the time they reach the other side they have been converted to nine-pound weights. The resulting imbalance will obviously spin the wheel like mad.

To those who wish to make one of these and use it, I offer the following economic advice. When you buy the weights, make sure you get six-pound weights and install them on the left side, as they cost only two-thirds as much as nine-pound weights.

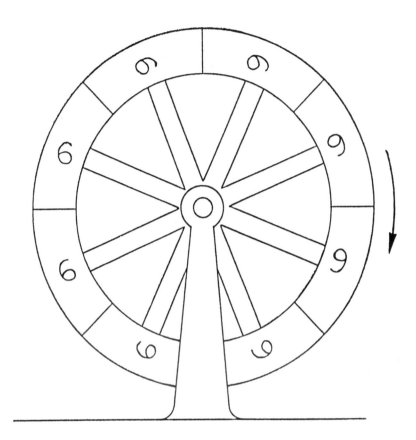

Perpetual-motion scheme

HUMOR AND CREATIVITY

This book is basically serious, but a little humor helps to keep us from taking ourselves *too* seriously. It is an interesting fact that most creative people are above average in the humor department also. I would guess that one reason for this is that both creativity and humor require the same basic intelligence. Another reason is that much humor is funny *because* it involves creative (but perhaps also silly) concepts, such as the six-gets-you-nine weighted wheel mentioned earlier.

Perhaps it is fortunate that most inventors have a humorous streak. It may help us keep our sanity in the often difficult and frustrating game of inventing.

PERPETUAL-MOTION PATENTS

By the rules of practice of the patent office, an invention must be workable in order to be granted a patent, and perpetual-motion machines can't work, so there can be no perpetual-motion patents. Not quite true.

The patent office makes special note of alleged perpetual-motion machines and says they are not patentable because of their lack of operability. An "inventor" can apply for a perpetual-motion patent, but the patent office will look at the application and promptly disallow it, or at least try to.

PATENT MODELS

Working models were required by the patent office as a part of all patent applications up until 1880. Some of these old patent models were very clever and sometimes beautiful, and they were always interesting. I recommend the Cooper-Hewitt Museum's book *American Enterprise* on patent models, listed in the bibliography. Also, if you are able to visit the Smithsonian Institution in Washington, D.C., spend some time in the fascinating patent-models section.

The patent office may still require a patent model if they feel it is necessary in a particular case. They always require a model in connection with perpetual-motion applications. The model prevents a lot of argument between the inventor or his or her attorney and the patent office. The model will either work or it won't. If it won't work it can't have a patent, period . . . well, usually.

FRAUDULENT PATENTS

In spite of all the safeguards to keep the patent process scientific, the patent office, to its embarrassment, sometimes unknowingly endorses frauds.

In 1981, I was a leader in a National Innovation Workshop, and Gerald Mossinghoff, then commissioner of the U.S. Patent Office, was the luncheon speaker. Speaking with him privately later, I reminded him of several perpetual-motion and other unworkable patents that had been granted, and I asked him how such things could happen. He said something like, "Sometimes we just get tired of fighting."

The heading "Fraudulent Patents" may be unfair and needs to be discussed. In some cases, an inventor knows he or she is perpetrating a fraud and tries to include the patent office in the list of victims. In other cases, a mistaken inventor believes in what he or she is trying to do and may succeed in convincing the patent office also.

In still other cases, an inventor is right and the establishment, including the patent office, is wrong. The patent office must be alert to this possibility. Protect us from a system that would refuse to examine and consider ideas that are counter to accepted thought. The patent office has a tough assignment, to separate deception and ignorance from progress. The distinction is often not clear-cut.

THE AMAZING MAGNETIC MOTOR

I have a copy of patent number 4,151,431, issued to Howard R. Johnson in 1979, titled "Permanent Magnet Motor." It is really a perpetual-motion machine, but the patent office couldn't give it that title because perpetual-motion machines are not patentable.

The abstract at the beginning of the patent reads:

The invention is directed to the method of utilizing the unpaired electron spins in ferro magnetic and other materials as a source of magnetic fields for producing power without any electron flow as occurs in normal conductors, and to permanent magnet motors for utilizing this method to produce a power source. In the practice of the invention the unpaired electron spins occurring within permanent magnets are utilized to produce a motive power source solely through the superconducting characteristics of a permanent magnet and the magnetic flux created by the magnets are controlled and concentrated to orient the magnetic forces generated in such a manner to do useful continuous work, such as the displacement of a rotor with respect to a stator. The timing and orientation of magnetic forces at the rotor and stator components produced by permanent magnets to produce a motor is accomplished with the proper geometrical relationship of these components.

United States Patent [19]

Johnson

[11]   **4,151,431**

[45]   **Apr. 24, 1979**

[54] PERMANENT MAGNET MOTOR

[76] Inventor: Howard R. Johnson, 3300 Mt. Hope
                 Rd., Grass Lake, Mich. 49240

[21] Appl. No.: 422,306

[22] Filed: Dec. 6, 1973

[51] Int. Cl.² H02K 41/00; H02N 11/00
[52] U.S. Cl. 310/12; 310/152
[58] Field of Search 24/DIG. 9; 415/DIG. 2;
       46/236; 273/118 A, 119 A, 120 A, 121 A, 122
       A, 123 A, 124, 125 A, 126 A, 130 A, 131 A, 131
       AD, 134 A, 135 A, 136 B, 137 AE, 138 A

[56]   **References Cited**

            U.S. PATENT DOCUMENTS

  4,074,153   2/1978   Baker et al. 310/12

Primary Examiner—Donovan F. Duggan
Attorney, Agent, or Firm—Beaman & Beaman

[57]   **ABSTRACT**

The invention is directed to the method of utilizing the unpaired electron spins in ferro magnetic and other materials as a source of magnetic fields for producing power without any electron flow as occurs in normal conductors, and to permanent magnet motors for utilizing this method to produce a power source. In the practice of the invention the unpaired electron spins occurring within permanent magnets are utilized to produce a motive power source solely through the superconducting characteristics of a permanent magnet and the magnetic flux created by the magnets are controlled and concentrated to orient the magnetic forces generated in such a manner to do useful continuous work, such as the displacement of a rotor with respect to a stator. The timing and orientation of magnetic forces at the rotor and stator components produced by permanent magnets to produce a motor is accomplished with the proper geometrical relationship of these components.

28 Claims, 10 Drawing Figures

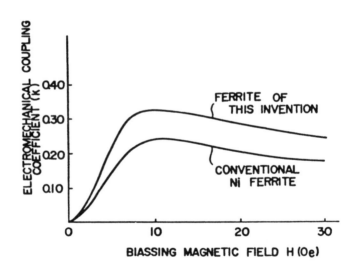

Permanent magnet motor patent

No power in, but lots of power out . . . not by my physics books! I have some experience in magnetics and have several patents in that field, so I tried to study and make sense out of the detailed body of

Mr. Johnson's patent. It made no sense to me, so I asked a fellow engineer who is an expert in magnetics to look at it. His verdict was "Complete gibberish."

In the spring of 1980, *Science & Mechanics* magazine published a long article on Johnson's invention. The writer of the article, who claimed to be a former research scientist, also claimed to have visited Johnson and to have personally seen the perpetual-motion motor run. He quoted Johnson as saying that the model was also successfully demonstrated to the patent office. It worked (?) and it was novel, so it got a patent, but only after appeal proceedings.

The one thought I have on how Johnson's models might have run, or appeared to run, is that they may have been frauds. The models were plenty large enough to hide a battery and an electric motor inside. If so, it wasn't the first time and won't be the last time that perpetual-motion machines have been provided with a common but hidden source of power during a workability demonstration.

It is now thirteen years after the "permanent magnet motor" patent was issued, and I have heard no more about this wonderful invention. We continue to search for energy sources. I just tried to contact Howard R. Johnson to get an update on his invention. According to the telephone company he has an unlisted number. Perhaps he received more publicity than he bargained for.

OTHER UNFULFILLED MIRACLES

In 1984, a national news story reported on the efforts of inventor Joseph Newman to get a patent on his perpetual-motion machine, but again, the inventor avoided the use of that term. Perpetual-motion machines won't work, so if you want it to work (or get a patent), don't call it that.

At that time Newman was fighting the patent appeals board in the U.S. District Court in Washington, D.C. Interestingly enough, he had support. Several competent scientists and companies had built hardware and tested Newman's concept, and they claimed it did develop an efficiency of well over 100 percent!

I don't know whether Mr. Newman ever got the patent, but it is now eight years later and we are still burning coal and oil for energy.

A short item titled "New Energy Source," in the September 1991 issue of *Sport Aviation* magazine, describes a new means of providing energy for airplane engines that is "being developed" by an Ohio firm. According to the magazine's source:

> The company has obtained a long list of patents covering every aspect of a system that involves a new way of accomplishing electrolysis of water . . . producing hydrogen rapidly enough to continuously operate an internal combustion engine. . . . The electrical power needed to break down the water into hydrogen and oxygen is said to be supplied by the engine's normal electrical system. . . . The device that induces on-demand electrolysis is about the size of an aircraft spark plug and, indeed, screws into a normal spark plug hole to perform its dual function of producing hydrogen and igniting it in the combustion chamber.

Glory be! The *Sport Aviation* article goes on to note that if this is all true, it is of course the answer to the world's energy problems. To their credit, they express doubt.

In 1959 Norman Dean received U.S. Patent number 2,886,976, titled "Systems for Converting Rotary Motion into Unidirectional Motion." The title of the patent is straightforward enough, but the invention isn't.

What Dean claimed, or at least what he seemed to claim, was repeal of one of Newton's laws of motion, "For every action there must be an equal and opposite reaction." Models were built and claims were made that they worked.

The "Dean Drive," as it was christened by pulp-science-magazine writers, would, if it *really* worked, make possible space flight without rocket engines. At least one writer expressed the opinion that flying saucers use the Dean Drive. That was thirty-three years ago. We are still using rockets. And the patent office still occasionally gets egg on its face.

When I was a kid my dad bought a gadget from a street vendor that makes an interesting story. The guy was selling "Atomic Condensers." That was ten years before the first atomic bomb, but even then the word *atomic* apparently had sales appeal.

The salesman had a junker of a car at the curb. After he succeeded in attracting a few suckers he would start the car engine. It would idle very roughly and perhaps die. Then he would open up the hood,

May 19, 1959   N. L. DEAN   2,886,976

SYSTEM FOR CONVERTING ROTARY MOTION INTO UNIDIRECTIONAL MOTION

Filed July 13, 1956   4 Sheets—Sheet 1

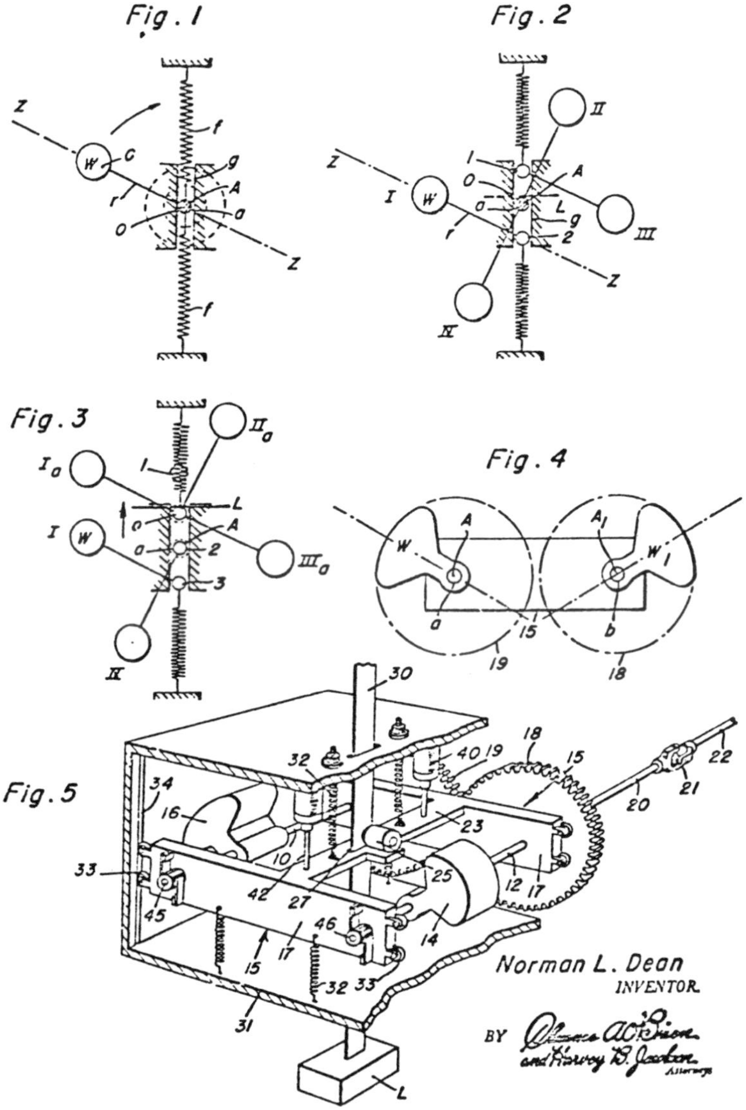

Fig. 1

Fig. 2

Fig. 3

Fig. 4

Fig. 5

Norman L. Dean
INVENTOR

BY

Dean patent

connect an Atomic Condenser between the spark coil and the high-tension lead, and restart the car. This time it would purr like a kitten.

I suspect he also threw a hidden switch when he restarted the car. At any rate, Dad was impressed, bought one, and put it on our car. The car ran fine without the device, and ran fine with it, so nothing was proven. Eventually he took the device off and gave it to me. I put it in a drawer in my shop and forgot it.

Years later, after I had obtained an engineering degree, I ran across the thing and became curious about it, so I ran some electrical tests and took it apart to see what it consisted of. Surprise! It was devoid of electrical function because it was wired straight through. It couldn't possibly affect the performance of a car ignition system.

Then I noticed a patent number on it, number 1,903,654, so I sent away for a copy of the patent, part of which is reproduced here. It is the shortest utility patent I have ever seen, two pages total.

The plot thickens. The patented device does specify a potentially useful electrical component (a spark gap), while the actual hardware does not have it! It is obvious that the item was marketed for the sucker trade, since its potential real value in improved performance is very limited and restricted to sick automobiles.

For the curious among you, another spark gap in series with the plugs could force the secondary voltage higher before it sparks, thereby increasing the current and the intensity of the spark at the plugs.

My guess at what happened is that the manufacturer observed the spark gap was costing him a few cents to put in, and he knew that most of the suckers wouldn't know the difference, so he left it out.

So we had a fraudulent device on the market, and the U.S. Patent Office was an unsuspecting party to that fraud because the patent they issued (on a different thing) was effectively being used as official government endorsement.

In my files is an advertisement for a "Magnetic Water Treatment Device." It is a small box that is clamped to the outside of a water pipe. The ad says the device contains a powerful permanent ceramic double-pole magnet. By the way, it is theoretically impossible to have a single-pole magnet, but I guess "double-pole" makes impressive advertising copy.

The ad goes on in a pseudoscientific fashion, being careful to avoid specific provable claims. It says the device is a conditioner, not a filter or purifier, which lets the company off the hook, since

April 11, 1933. H. G. OESTRICHER 1,903,654

VOLTAGE AMPLIFIER

Filed Oct. 11, 1932

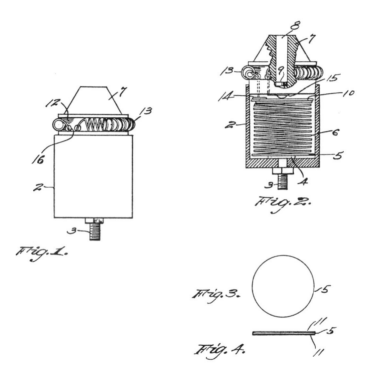

Fig. 1.
Fig. 2.
Fig. 3.
Fig. 4.

INVENTOR,
H. G. Oestricher;
BY F. E. Maynard
ATTORNEY

Voltage amplifier patent

"conditioning" is a rather vague action in the case of pure water. The ad doesn't mention any patent. I wonder if they tried to get one.

The ad talks about the device's magnetic field affecting the scale-forming minerals in the water. That is interesting, since these minerals are nonmagnetic. Of similar interest is the ad's failure to recommend the device for use on nonmagnetic pipes only. Galvanized steel pipes would shunt the magnetic flux and keep it from getting into the water. I guess since the magnetism doesn't do anything to the water anyway, it will work just as well on magnetic pipes.

Caveat emptor, let the buyer beware.

7

Basic Physics, Anyone?

The art of inventing is so dependent upon a basic understanding of simple physics that I feel the subject is worth a chapter here.

I suspect you are a diverse readership. Some of you will know all of the following and much, much more, but for those who don't, please pay attention. Of all the fields of knowledge, I find physics to be the most fascinating, but I realize that science in general, and physics in particular, scares off a lot of people. Let me see if I can make the subject even slightly as interesting as it really is.

WORK, ENERGY, AND POWER

Many people confuse these concepts. Work and energy are different forms of the same thing. In order for a perfect machine to do a given amount of work it must consume that same amount of energy, in the form of fuel or electricity. Real machines consume more, as we will see.

Engineers measure work in the unit of "foot-pounds," or its metric equivalent. If you raised ten pounds of something up a distance of two feet, you did ten times two or twenty foot-pounds of work.

There are 3,087 foot-pounds in a calorie, so a hundred-pound woman would have to walk up a 30.87-foot-high flight of stairs, or the equivalent, to burn off one of those dreaded calories, *if* she was a 100 percent efficient machine. We are far from perfect machines, however, so it is actually much easier than that to get rid of excess calories by exercise.

Electricity is another form of energy. A kilowatt-hour equals 2,655,000 foot-pounds, so it could do a great deal of work. But what is a kilowatt-hour, aside from something that appears on our electric bills? The watt is a unit of power. A thousand watts for one hour is one kilowatt-hour of energy. You can burn ten hundred-watt light bulbs for an hour and your electric meter will show one kilowatt-hour more.

So a watt is a measure of power, but what is power? Power is the rate of doing work or the rate of using energy. Whether the woman walked up the stairs or ran up the stairs (assuming her efficiency walking and running are the same), she will burn the same one calorie, but she will obviously burn it faster while she is running. The power she puts out is also greater when she is running.

A powerful person, when compared to a weak person, or a powerful engine or motor compared to a weak one, is able to do more work in the same length of time or do the same work in less time. Power involves time; energy and work do not.

A strong man is not necessarily a powerful man. In technical use, the two words *strong* and *powerful* mean different things. A strong man can lift a heavy weight; a powerful man can lift it rapidly; an energetic man can keep on lifting it longer. Unfortunately, in everyday language we tend to broaden and corrupt the specific meanings of technical words.

We measure power in units of horsepower or watts. A horsepower is 550 foot-pounds of work per second and is also equal to 746 watts. So a perfectly efficient one-horsepower motor (let me know if you can find one) would use 746 watts of electricity continuously, or nearly three-quarters of a kilowatt-hour in an hour, while developing one horsepower.

Considering the real efficiency of motors, a one-horsepower motor will actually require about one kilowatt of electricity when working at its one-horsepower-rated load. A kilowatt-hour costs only about ten cents in many parts of the United States. Try to rent a horse for an hour for ten cents. Actually, a horse can put out a lot more than one horsepower for a short time. One horsepower is nearer to the power level a plodding horse could work all day long.

Let's look at another example. Assume we have a creek with a ten-foot-high waterfall in which fifty-five pounds of water is falling every second. Conveniently, 10 times 55 equals 550 foot-pounds of work per second, or exactly one horsepower. Weren't we lucky that it came out so even!

Also assume we are in an unreal world where everything is perfectly efficient. We could put a perfect hydraulic turbine in the waterfall and get one horsepower of mechanical power out of it. We could connect a perfect electric generator to the turbine and generate 746 watts of electricity with it. We could connect that power to a perfect electric motor and again have one mechanical horsepower.

We could also connect a perfect hydraulic pump to the motor and pump all of the fifty-five pounds of water we used in one second back up to the top of the waterfall in one second. That one second's worth of water at the top has 10 feet times 55 pounds or 550 foot-pounds of potential energy, just as it had before we did all those things.

Not that we would want to go around in a circle like that, but it illustrates some of the ways we can convert work or energy and power into different forms. And not that we *could* do it without loss. This was a make-believe exercise. To make it really work we would need perpetual motion, which can't exist.

In the real world almost everything is inefficient to some degree. To find the percentage of the original fifty-five pounds of water that we could actually pump back up to the top in one second, we would have to multiply in turn by the true efficiencies of the turbine, the generator, the motor, and the pump, and also by the efficiencies of the pipes in which the water ran and the wires in which the electricity was conducted. We would be doing very well if we could actually pump half of the water back up each second.

For those of you who never thought about any of this before, it was probably confusing. In these short paragraphs you have been exposed to much of a quarter of high school physics. Like all subjects, physics is more easily learned one step at a time: note how simple and easy-to-understand each step is by itself.

FORCE AND DISTANCE

A brief look at forces will also be very useful to potential inventors. We know that with a lever we can lift a greater weight than we can lift without the lever. Levers can multiply force, but isn't that getting something for nothing? Isn't it demonstrating greater than 100 percent efficiency?

No, because it costs us something to get more force than we start with by means of a lever. We are trading distance for force. With a certain lever we might lift a thousand pounds where we would normally be able to lift only a hundred pounds; but our end of that lever must move ten inches if the thousand-pound weight is raised one inch at the other end.

The energy expended at one end of the lever always equals the work done at the other end. In this case ten inches times a hundred pounds equals a thousand inch-pounds; and one inch times a thousand pounds also equals a thousand inch-pounds of work. We got what we paid for, and no more. Actually, we will get a little less, because of the friction in the lever fulcrum bearing.

VOLTAGE AND CURRENT

Other units of electricity are also interesting to look at. Volts are really units of force or pressure, a little bit like the height of our waterfall. Amperes are units of current, and they can be compared to the flow of water over the falls. The bigger the flow or current of electricity, the more amperes we have. If we multiply the current in amperes by the voltage, we get watts, a unit of power, as we have already discussed.

Transformers, such as those used to power electric trains, and the big ones on the power poles in the street, are neat devices for juggling current and voltage, much as a lever juggles force and distance.

If our son's toy-train transformer is plugged into 120 volts and uses a current of 1 ampere, it is consuming 120 volts times 1 ampere, or 120 watts of power. That transformer was designed for an output of about 10 volts, to drive the toy train. In a perfect world, it could deliver 12 amperes to the train, because the output power would equal the input power, and 10 volts times 12 amperes equals our original 120 watts input.

Transformers are very efficient devices, but they can't be perfect, so we would actually get more like 11 amperes at 10 volts instead of the ideal 12 amperes. We got a little less power out than we put in, but we obtained the lower voltage and higher current that we wanted.

If you have a gas furnace, or an oil furnace, it probably has an ignition transformer that is used when the furnace starts. In this case the input voltage is still 120, but the output is raised to around 20,000 volts to provide a hot spark to ignite the fuel. From 120 to 20,000—Wow!

Did we get something for nothing? No—the output current is only a few thousandths of an ampere, while the input current was nearly two hundred times as much. We can't get quite as much power out as we put in.

There is no free lunch. More to the point, there is no perpetual motion. The intelligent inventor uses the laws of physics to do new and useful things and doesn't beat his or her brains out trying to disprove these natural laws.

HEAT

Heat is called the lowest form of energy. By that scientists mean that all other forms of energy have a tendency to turn into heat (flow "down" to heat, the lowest form). For instance, the friction in a bearing causes the bearing to get warm. Part of the mechanical energy of rotation is converted to heat because of the inefficiency of the bearing.

When the light in a room hits the walls it is partly absorbed and turned into heat. White walls reflect most of the light, but dark walls absorb most of it, so dark walls will actually be a little warmer to the touch than light-colored walls.

Incandescent lights are very inefficient sources of light. Most of their input electrical power is converted into infrared radiation (radiant heat) and into convected and conducted heat. (Fluorescent lights and LEDs are more efficient.)

So a light bulb is an inefficient source of *light* but a very efficient source of *heat*. In fact, if we close the blinds and doors so no light gets out of the room, a light bulb becomes a 100 percent efficient *heater*! All of the light and infrared radiation that it puts out is absorbed by the walls and other things in the room and converted to heat.

Sound is also absorbed by surfaces it contacts, and by the air, warming them slightly. If you have a boom box putting out ten watts, and if your house is so well insulated that no sound gets outside, your house will be warmed by that sound by the same amount that a ten-watt heater would warm it.

Even a refrigerator is a 100 percent efficient source of room heat. It cools its own box and the food in it, but the heat it takes out of the box is dumped into the room by the radiator coils in the back, along with the heat generated directly by the inefficiency of the motor and the compressor.

I once heard of a person who wanted to leave the refrigerator door open on hot days to cool the kitchen. Of course, it doesn't work that way. The more the refrigerator runs the more it heats the room, and with the door open it would run constantly. The "cold" that would spill out of the open refrigerator would not nearly keep up with the heat the refrigerator was producing.

A fan in a room is also a 100 percent efficient heater. It will make you feel cooler because the circulating air will improve the heat transfer from your body, but all of the moving air it stirs up is eventually stopped by friction, and all the energy in the air due to its motion is converted to heat. As in other machines, the energy lost in the motor and at the fan blades due to their inefficiency is directly converted to heat.

The energy "lost" due to the inefficiency of machines doesn't just disappear, it is always and totally turned into heat. Earlier I mentioned that almost everything is inefficient to some degree. The exception, as we can now understand, is the electric heater.

It takes work to cut vegetables. Where does that work go? The cut vegetables have no more calories than the whole ones did. The knife experienced friction in the cutting process. That friction turned all of the cook's cutting work into heat in the vegetables and knife, and ultimately in the room.

The high efficiency of heaters is handy when we need heat, but the fact that heat is the "low man" on the energy totem pole is very troublesome when we want mechanical or electrical energy.

Steam engines and turbines, internal-combustion engines, and gas turbines are all called "heat engines" because we are turning the chemical energy of fuel into heat by burning it, then converting the heat into mechanical motion.

This higher form of energy requires that we push the lower form (heat) uphill, so to speak. The result is very poor efficiency. Even the theoretical efficiency of heat engines is very low. In other words, a "perfect" heat machine would still have poor efficiency.

The thermodynamics that explain all of this in scientific terms were worked out by French engineer Sadi Carnot in 1824, when he was only twenty-eight years of age. Ah, to be brilliant.

BEATING CARNOT

The inventors who try to invent perpetual-motion machines are, usually unknowingly, trying to prove Monsieur Carnot wrong. It can't be done, because he was right. But he can be bypassed!

If we use heat engines to get from fuel energy to mechanical energy, as we always have, we must accept the brilliant Frenchman's low efficiencies. But why use heat engines? Nature doesn't! All animals convert the chemical energy of fuel (food) directly into mechanical energy without converting it into heat first! Plants that have any motion of their own also do it. Trees that push a thousand pounds of their own weight up a hundred feet or so don't do so at high temperatures. Muscles and plants can be very efficient because they are not heat engines.

Animal-muscle energy conversion is nonthermal and therefore not constrained by Carnot's laws. But, as yet, we don't understand at all well what is going on in animal muscles.

I briefly worked on the problem of inventing a machine that would convert fuel directly to motion, but I didn't get far. It is a very tough assignment and will take better brains than mine. Why bother? Because nonthermal energy conversion will not only be more efficient (thereby helping to conserve our limited fuel supplies), it will probably be environmentally cleaner and may also be cheaper, smaller, lighter, and safer.

I would love to spend the rest of the book discussing this fascinating challenge, but we have other things to talk about. Sometime in the future when you hear that some inventor, engineer, or scientist has successfully developed a nonheat engine, remember you heard about the goal here first.

THE ELECTROMAGNETIC SPECTRUM

Did you know that rainbows, cosmic rays, heat, ultraviolet light, and radio waves are all the same thing?

If we bring a magnetic compass near a wire carrying a current, the compass will deflect, because there is a magnetic field at right angles to the electric field. Likewise, a permanent magnet has an electric field surrounding it at right angles to its magnetic field. Electricity and magnetism can't exist without each other (how romantic).

When we start alternating an electric current or varying a magnetic field, the magnetism and the electricity start talking to each other. (It was a pretty boring marriage up to this point.) This interaction is the principle behind all electric motors, transformers, speakers, etc., where the electricity is flowing in wire coils and the magnetism is in iron cores.

Neither electricity nor magnetism needs to have the wire or iron, however. Both can flow through space as waves, and do, no matter what the frequency of the "alternating current." But in all cases this radiation will consist of both an electric field and a magnetic field at right angles to it, hence the name "electromagnetic radiation."

Now look at the chart on the next page. About the lowest point of interest is our alternating-current power, where the frequency is only sixty cycles per second (fifty in Europe). Saying "cycles per second" is a no-no nowadays, however. One cycle per second is called one "hertz," in honor of Heinrich Hertz, who generated the first man-made radio waves.

By the way, for those unfamiliar with the system for denoting large numbers, "ten to the sixth" (written as 10^6) means a one with six zeros after it, or one million.

The lower the frequency, the larger the "transmitter" must be to radiate much electromagnetic energy. At sixty hertz the amount of radiation is very small, but enough that people are beginning to wonder about the effects of living under a high-voltage power line.

The AM radio broadcasting band is from a frequency of half a million to one and a half million hertz. You will note that as the frequency gets higher, radio waves turn into infrared, then into visible light waves. We talk about the colors of the rainbow as "the spectrum," but actually the visible spectrum is a very, very small part of the total electromagnetic spectrum.

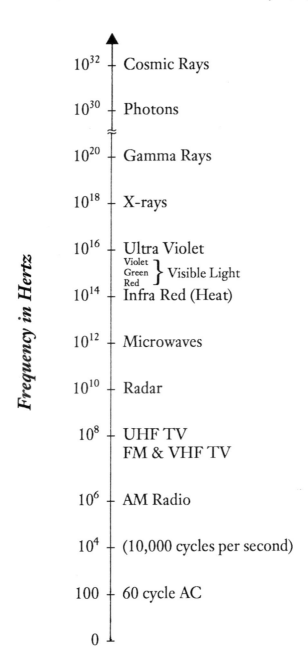

The electromagnetic spectrum

As the frequency continues to increase, light waves turn into ultraviolet and on into X-rays, and the X-rays turn into nuclear radiation and then into cosmic rays.

Electromagnetic radiation is always the same combination of electricity and magnetism. We give different parts of the spectrum different names, since the various frequencies affect our human senses in different ways and we have different uses for them. A radio transmitter doesn't look much like a light bulb, which isn't much like an X-ray machine, but they are all putting out different frequencies of that same electromagnetic radiation. The galaxies in space radiate pretty much the entire spectrum. Now isn't that fascinating?

8

Inventor Groups

CLUBS AND SOCIETIES

Inventor organizations have a lot to offer inventors, especially beginners in the game. I have been a member of the American Society of Inventors, headquartered in Pennsylvania, and I belonged to the Inventors Association of Washington until it disbanded a few years ago.

I am currently a member of the Minnesota Inventors Congress (P.O. Box 71, Redwood Falls, Minnesota 56483-0071), a very active and effective nonprofit corporation that mails out useful literature. You may be able to find a local group to associate with, but getting on the mailing list of a large club with a good newsletter is worthwhile, even if you can't attend their meetings.

Most inventor-club meetings are planned around selected speakers on subjects of value to the membership. The speakers may be patent attorneys, investors, successful inventors, manufacturers, marketing people, and so forth.

INVENTION FAIRS

Inventor clubs, chambers of commerce, shopping malls, science centers, and similar organizations sometimes put on invention fairs.

The fair may or may not be a part of a conference or symposium where valuable talks on invention are given, but here I wish to address just the invention fair.

Sometimes booth space is free, but usually an inventor is charged a rental fee for a booth to show off his or her invention for the one or two days of the fair. The management of the invention fair will frequently promise that manufacturers will be invited to the fair, where they will have an opportunity to talk to the inventors of products they may be interested in manufacturing.

In my opinion, few if any serious manufacturers come to such invention fairs. Most manufacturers are not looking for new products, and if they are, the inventors will seek them out directly. I am not aware of any licensing agreement or sale of a patent that resulted directly from a manufacturer being introduced to a new product at an invention fair.

Invention fairs seem to me to serve several purposes, but selling the rights to inventions is not one of them. Some invention fairs allow small-scale retail selling of new products from the booths. The public comes chiefly to be entertained, and often is, while the inventors get an ego boost from showing their inventions.

Often an inventor will try to get what he or she considers a marketing survey of an invention by asking people who stop at the booth what they think of it and if they would buy one when it is put on the market. Such a survey is invariably very biased, misleading, and can be financially ruinous to the inventor. The problem is that people don't like to hurt other people's feelings. Very few people will say they think the invention is no good.

Many will say they would want to buy one if it were available, but making an idle statement to a stranger you will never see again is entirely different from putting cash on the line for a product. The difficulties in getting reliable market surveys will be further addressed in Chapter 20.

NATIONAL INNOVATION WORKSHOPS

For many years the Small Business Administration and the National Bureau of Standards have been sponsoring and subsidizing a series of innovation workshops in major cities. These are sometimes held on the campuses of universities and sometimes in other facilities.

I have participated in a number of these workshops and consider them to be a good deal for inventors. Because the government picks up much of the tab, it costs little for an inventor to register. (I'm not sure this program is worth its cost to the taxpayers, however.)

Experts in many fields relating to inventing and marketing are invited to lead the workshops, and Small Business Administration people speak and make available various government pamphlets on subjects of interest to inventors.

INNOVATION ASSESSMENT CENTERS

The U.S. government also sponsors a few "Innovation Assessment Centers" around the country for the purpose of promoting small business. These are usually associated with some university. For a small fee the center will evaluate inventions from private inventors or small companies.

The evaluators are professors from the university or other experts. The evaluators work for free (I know, because I am one of them), which helps hold down the cost to the inventor.

A formal evaluation procedure is followed, which addresses all aspects of the invention's chances of success in the marketplace. This evaluation is far better than one the inventor would be able to make alone, but it is not perfect.

I'm sure these centers have given discouraging evaluations to some inventions that were good, and they have likewise given good marks to losers. No person or group can accurately predict whether the marketplace will accept or reject inventions; they can only make educated guesses.

I had the Innovation Assessment Center based in Seattle evaluate one of my own inventions. They gave me some useful advice, but they completely missed the fact that the invention was already being produced and sold by five major companies. I discovered that fact later on my own.

9

Liability

In this age of legal suits for all kinds of damages, the possibility of suits against inventors is of concern. I have studied the reports of suits in general, and I have asked a number of patent attorneys if they were aware of any suits against inventors that involved physical injury or death. So far the answers have been negative.

From my career in engineering management at Boeing, I know that every time an airplane goes down for any reason, lawsuits spring up like wildfire. Attorneys contact the survivors and the relatives of the deceased and offer joint-action suits in their behalf.

The airplane company is sued, and the airplane engine company, the instrumentation companies, the air-traffic control authority, the FCC, the FAA, perhaps the U.S. Weather Bureau, and any other organization that has money and might possibly be connected to the crash. These suits go on for years, and large corporations have billions of dollars at stake in them. The net result is that the attorneys get rich and most of us get poorer through higher airline fares.

Now that I have that off my chest, how about inventors? In my opinion, the chief reason that inventors are seldom sued is that inventors are usually poor. It is a waste of time and money to sue someone who couldn't pay if he or she loses.

68

However, we inventors *could* be sued; therefore it behooves us to protect ourselves to the best of our ability. In this regard, I think it could help to write thorough and accurate disclaimers into our patent applications.

In 1984 I was awarded patent number 4,431,182 on "Human Free-Flight Amusement Devices." Although safe if properly designed, built, maintained, and used, this invention had some potential for injuring and killing people. We have hundreds of inventions in this category, including airplanes, automobiles, boats, skis, bicycles, guns, and trampolines. In view of the current popularity of liability suits, I included in the patent an extensive section on proper design and safety.

I made presentations to five companies in the amusement-rides business in an attempt to sell or license this patent. In all cases they showed considerable interest during and immediately after the presentations, but in all cases they later decided they were not interested. I am quite sure that the invention was rejected primarily because of its potential for generating liability suits.

The following quotation from a speech by Lee Iacocca expresses my concerns very well: "A small company in Virginia that made driving aids for handicapped people went out of business because it couldn't afford the liability insurance. Too risky. Hardly anyone makes gymnastics or hockey equipment anymore. Too risky. We've virtually stopped making light aircraft in this country; the biggest cost is the product liability. Too risky. One day, we're going to wake up and say, 'The hell with it—competing is just too risky!' Why even try to build a better mousetrap? Let somebody else do it—and then sue him."

10

Creativity Is a Personal Thing

Creative people tend to find problems or see challenges where other people see no problem or need for a change. Creative people with an inventive bent tend to take longer and be more thorough in analyzing and formulating a problem, in collecting data, in studying, scrutinizing, and searching for contradictions. These people are naturally interested in such matters, while most people would consider them boring.

Creativity is an ego-satisfying thing. A person working on a creative problem is apt to feel that he or she is the only one who can do that job properly. Vandyke, the seventeenth-century Flemish painter, said, "This is my work, my blessing, of all who live, I am the one by whom this work can best be done in the right way."

That attitude may seem conceited, but that kind of self-confidence is necessary before we can create. We must believe that we can do something that has never been done before, or we would never try. Henry Ford said, "If you think you can, you are probably right. If you think you can't, you are definitely right." Believing you will fail is a self-fulfilling prophecy.

The Christian ethic of humility has its place, but pride in one's work is essential in inventing. Inventors want to show off their inventions. That is natural, necessary, and good. Ego provides drive, which promotes progress.

We all need to be liked and loved. We also need to be admired. Creative accomplishments earn admiration for us. Desire for that admiration, whether conscious or subconscious, is a source of drive to make us produce. Inventing is hard work that often requires sacrifices. We don't work unless we have things to gain by it. Personal satisfaction in the work itself is one such thing, recognition is another.

Beginner inventors are apt to think they are inventing in order to get rich. Desire and need for money is certainly a driving force, but as we shall learn from this book, inventors seldom make money. Therefore, the nonmonetary rewards of inventing are very important.

Another indication that creativity is a personal thing comes from the sources of our problems. Routine problems and jobs are usually assigned to us by our parents, our teachers, our bosses, and our spouses. Creative problems, however, we almost always assign to ourselves. No one else is apt to tell you what to invent, because no one outside of you knows your personal creative inspirations.

Initially, creative thoughts aren't even clear to their creator. Buckminster Fuller said, "A thought is often only dimly and confusedly conceived at the moment of first awareness of that as yet vaguely describable thinking activity."

Irving Langmuir, Nobel Prize winner in chemistry, wrote, "Only a small part of scientific progress has resulted from planned search for specific objectives. A much more important part has been made possible by the freedom of the individual to follow his own curiosity."

THE UNCONSCIOUS MIND

You have probably heard of the subconscious or unconscious mind, and perhaps think it exists in theory only. It is real all right, but remains a bit mysterious because of our limited access to it.

You have also probably read of the left-brain/right-brain theories. I don't know offhand if the subconscious is supposed to be in the left half of the brain, the right half, or both, but it doesn't matter to me. This book is concerned with practical use of the brain, not theories on how it is organized. Understanding how the subconscious mind works, however, is important to us because it is very useful to inventors. If we play by its rules we can make it even more useful.

Studies and tests have shown that most or perhaps all of our creative output comes from the subconscious mind. We are not normally aware of this in working simple problems because the problem is solved essentially instantaneously, and we assume the conscious mind did the job.

In the case of more difficult creative problems, however, the solution takes longer and the involvement of the subconscious becomes easier to visualize. The conscious mind formulates the statement of the problem and the ground rules by which it is to be solved, and passes these on to the subconscious.

The subconscious plugs away at the problem, searching for answers. When it gets one it passes it back to the conscious mind, which evaluates and either accepts or rejects it. If the idea is rejected but the person still has an interest in the problem, the subconscious goes back to work, searching for another answer.

This whole activity reminds me a lot of the operation of a computer spell-checker suggesting possible correctly spelled words for the typist. If the needed word is simple and the input word was close to correct, the computer's answer is immediate. If the problem is more difficult the computer takes longer before it comes up with an answer. If its first answer is rejected it continues to search its memory and continues to make suggestions. Whether computer memory or subconscious mind, the operator or conscious mind is in control and makes the judgments and the decisions.

The conscious mind and the subconscious can operate at the same time on the same problem, or independently. When we sleep, it is the conscious mind that sleeps. The subconscious mind continues to work, at least part of the night.

Years ago I would sometimes be working on a creative problem late in the evening and not getting anywhere. I would decide to go to bed and worry about the problem the next day. On a number of occasions, as soon as I woke up in the morning I immediately saw the answer I was looking for. My reaction at that time was that the tired mind in the evening couldn't solve problems that were easy for the rested mind in the morning. I now know that what really happened was that my subconscious mind continued to work the problem while I was asleep, found an answer, and gave it to the conscious mind as soon as it awakened.

Another indication of the working of the subconscious during the night is seen when the subconscious doesn't wait for us to wake up,

The "Eureka!" light turns on when the solution is presented to the conscious by the subconscious mind. *Photo by Marianne Reynolds.*

but presents the answer to us in a dream. The subconscious independently works the problems we assign to it and appears to make its answers known as soon as it can get an audience with the conscious. That transfer of knowledge can occur at any time. It may occur immediately, in a dream, in the morning, in a week, a month, or years later, depending on the difficulty of the problem.

Psychiatrist Alexander Reid Martin felt that these insights from the subconscious were more apt to occur during periods of relaxation rather than periods of peak activity. This makes sense—the subconscious waits until the conscious isn't so busy before it interrupts.

The passing of the answer from the subconscious to the conscious is commonly mistaken as the moment when the conscious solved the problem, the "Eureka!" (I have found it) moment, the moment when the light bulb turns on over our heads. But the conscious didn't solve it—the subconscious did. The subconscious lets the conscious take credit and doesn't feel hurt. Like a computer, the subconscious doesn't appear to have emotions.

A computer works only by its own rules. So does the subconscious mind. You can't rush it; neither threats nor incentives do any good.

But if you are interested in the problem, it will work faithfully. I like to call our subconscious mind our "computer mind."

We tend to think of creative thoughts as coming "out of the blue." Not so. The experts in the field look on most creative thinking as the process of relating known facts to each other in unconventional ways. Like the computer, the subconscious mind cannot generate new knowledge out of nothing, but it can process what it has been given in remarkable ways to create new and useful combinations.

This fact makes clear the extreme importance of education and knowledge to inventors. If the brain has nothing to work on, it can't work. The more information in storage, the more creative the output.

The other side of that coin is misinformation. If you learn things that aren't true, your creative output is apt to contain many solutions that won't work. Again the computer analogy, "If garbage goes in, garbage comes out."

I use my subconscious mind consciously, if you will allow the play on words. If I make reasonable requests of it, my computer mind gives me good answers more than 90 percent of the time.

Since the subconscious is not able to spot misinformation, it is very important to give it an accurate statement of the problem and to make sure that no false assumptions or unrecognized assumptions are being passed on to it. Assumptions are fine if they are recognized as such, tried, and then rejected if they are found wanting.

In my invention classes I use a little puzzle that illustrates the danger of unrecognized assumptions. I ask each student to assume he or she has six matchsticks and invite him or her to make four equilateral triangles from the six sticks, where each leg of each triangle must be the full length of a matchstick.

Most students are unable to work the problem. They draw triangles on a piece of paper, but it doesn't come out. They fail because they have subconsciously assumed that the problem is two-dimensional. It isn't. The solution is a three-dimensional three-sided pyramid, or a regular tetrahedron.

Experienced and cautious persons will ask themselves whether the solution is two- or three-dimensional, and try both options. The failure of those who can't work the problem is not due to the assumption of two dimensions, but due to the fact that the existence of an assumption isn't recognized, so there is no escaping from it.

If the subconscious has sufficient knowledge to solve a problem, it eventually will, assuming that it is "creative" enough. But here we see that a person with great creativity must have both great knowledge in the fields of interest and great knowledge-processing ability.

You will note that I wrote *fields*, not field. The creative relating of facts to each other may, and frequently does, involve combining facts or knowledge from different and normally unrelated fields. The broader the education, experience, wisdom, and knowledge of an inventor the better.

In cases where the subconscious takes a very long time but finally solves a problem given to it, the person probably did not have enough knowledge to solve the problem initially, but acquired an essential missing fact in the interim. My computer works as far as it can on a problem, then patiently waits. When I give it the final bit of information it needs, it spits out the answer I want. So it is with the mind.

Most creative problems have more than one answer. Therefore, if the first answer your subconscious presents doesn't suit you completely, reject it, at least temporarily. There will probably be other suggestions forthcoming. In fact, even if the first answer looks very good, ask the subconscious to try some more anyway. It will often come up with an even better one.

As we will see in Chapter 23, it is important to discuss in your patent application all of the possible ways of implementing your invention you can think of in order to get the strongest possible protection. Therefore, even if the first answer is perfect, you need all the other answers the subconscious can come up with, to include in the patent.

AGE AND CREATIVITY

It has been pointed out that the more experience and knowledge, the better the creativity. This would seem to imply that older people should be more creative than younger people. Let's talk about that.

If older people could remember all of the important knowledge and wisdom they have learned over the years, I believe they would be more creative. There is no question that we tend to start forgetting some things as we pass middle age, however. What we remember depends on our interests. I have long ago forgotten such things as birthdays, people's names, and movie plots, but I can still tell you the speed of light, Ohm's law, and the number of feet in a mile.

Infants are said by some writers to have more creativity than adults. This "creativity" must be of the willingness-to-try-weird-combinations-and-dumb-things type, since the infant has almost no knowledge base to work from. Therefore, an infant's creative output is of little value, except for its own amusement and possibly that of its parents.

The value of the child's creative output to the parents varies from positive, in bragging to others about what the brilliant child did, to negative, when the creative effort was smearing strained beets on the wall.

This inherent human creativity tends to suffer as the child matures, however. Mom says, "No! Jimmy, don't smear your beets." The neighbor says, "Stop picking my flowers." The teacher says, "James, you must not mark Shirley's dress with that felt pen." The college professor says, "Smith, your method might be right, but I want the one in the textbook." The boss says, "Jim, your conclusions are creative, but they are not what we want."

These rebuffs are painful! We avoid pain by eliminating its sources. If creative actions get us into trouble, scratch creativity.

Some of us with strong individuality and a tolerance for pain persist in creative thought and actions and remain inventive throughout our lives. For these select individuals this book is written.

The decay in creativity due to the conformist pressures of society occurs, when it does, mostly in our early years. Beyond the mid-twenties, our creativity will probably remain fairly constant; so, in my opinion, aging per se is not a major factor in declining creativity.

We have been talking about creativity, however, not about inventing. There is no question that young inventors are far more prolific and develop and invent far more useful things than do older inventors—"young" meaning ages twenty to forty.

The reasons for this decline are waning ambition, energy, drive, and incentive, not waning creativity. The older inventor has probably already become famous, if he or she is going to be. He or she has either made a million dollars or has given up on being rich. He or she has tangled with the frustrating problems and pressures of inventing for years and would just as soon relax a bit now. Older inventors usually continue to tinker, but they seldom produce much of real significance.

Almost all inventors and scientists fit this mold. Edison invented the incandescent lamp, the nickel/iron storage battery (which led to our nickel/cadmium battery), the phonograph, the carbon telephone transmitter, and made contributions to the telegraph and to motion pictures all when he was young. Most of his later efforts came to naught.

Nikola Tesla invented alternating-current systems—including polyphase AC motors, alternators, and transformers—and made many basic contributions to wireless communication and robotics when he was young. Later he played around with artificial lightning and made wild claims about wireless power transmission that never materialized.

Einstein formulated his theories of relativity and discovered the interchangeability between matter and energy, which led to the nuclear age, when he was young. In his later years he dreamed and wrote abstruse equations but produced little of significance.

Looking at my own humble case, I earned most of my patents before age forty-three. My earlier efforts were also more important and more difficult than my later ones. I am now seventy-three. I still create and invent, not to make money, but because it is one of the most fascinating and satisfying hobbies there is.

11

Developing Our Creativity

Creativity has been a popular subject in personal development circles in recent years. We hear about creative financing, creative macrame, and other endeavors, but our interest in creativity here is in its application to inventing.

There are many factors involved in creative thinking. Some of them, like native intelligence, natural curiosity, and observation, may tend to be fixed, but some of the aspects of creativity are definitely improvable.

As we saw in the previous chapter, our practical creativity is limited by our knowledge, which we can always expand. Inventors should get a broad education. We may stop going to school, but we should never stop learning. It has been said, with much truth, that graduation should be more nearly the beginning of our education than the end. We never graduate from the school of life.

We can also improve our effective creativity by removing the barriers to creative thinking, so that our creative output more nearly matches our natural ability. That process and its value will be the subject of this and the next several chapters.

HABITS

Habits, traditions, conformity, and culture frequently stifle creativity. Some of the reasons for this are discussed in Chapter 3, "The World Dislikes Inventors."

We all have personal habit patterns. We are conscious of some of these, but we do not know we have many of them. As an example, conduct this little test while you read this.

Fold your two hands naturally, interlacing the fingers. Now observe and remember which thumb you put on top, the right or the left one. Separate your hands and again fold them naturally without thinking about it. Did you put the same thumb on top? Ninety-five percent of all people will. That is a habit most of us develop.

This time, fold your hands the other way, putting the other thumb on top. Did it take some conscious effort? How does it feel—strange, uncomfortable, even a little upsetting? For many people it will, because it is wrong. That is not the way they fold their hands.

It is improbable that the way you fold your hands will affect any invention you may work on, but other equally obscure habits you may have could well affect your creative efforts. Note that it will probably not be the habit itself that may affect your thinking and inventions, but the fact that you don't realize you have the habit.

As a hypothetical (and unrealistic) example, assume you are left-handed but don't know that most people are different than you are in that respect. If you are trying to invent something having to do with writing or drawing, or putting on clothes, or using tools, you will probably design for the minority rather than the majority, and might financially regret it.

TRADITIONS

It strikes me that "habits" are personal, but "traditions" are societal, group, or cultural habits. At any rate, traditions can have the same stifling effect on our creative thinking as our personal habits do.

DON'T BE A CATERPILLAR

Jean-Henri Fabre, the great nineteenth-century French naturalist, was studying tent caterpillars on a tree in his yard one day. He observed that their tradition was to follow the leader, in one long single file, and he decided to conduct an experiment. He broke the train of caterpillars and led some of them onto the rim of a flowerpot until there was a complete circle all the way around.

They were following each other as before, but there was now no leader in the endless circle. They continued to march, around and around, for hours. Meanwhile, Fabre got tired of watching them and went to bed.

When he got up the next morning and checked on his experiment, the caterpillars were still marching around the pot rim. Fabre figured they were probably very hungry by this time. Therefore, he supported the pot next to the branch where he originally found them so that any one of them could have switched over to the tree and led the others, and they could have eaten to their hearts' content.

But no . . . the tradition of following the leader was too strong; they continued to march in the circle until they died of starvation and exhaustion.

Nonthinking humans have also been known to go around in circles.

DICE ARE CUBICAL

There are other shapes in which dice might be made and still be useful in games of chance, but the tradition is strong—dice are cubical.

Let us assume we would like to invent dice (more than one die) that are not cubical. Before we can do that intelligently we need to establish the requirements for dice.

When I give this problem to my classes, the students rapidly decide that dice must be of regular shape so there is an equal probability of any side coming up; they must stop rolling in definite positions with a number up; and they must have six numbers, because our dice games have been designed for six-number dice.

I ask my classes if spherical dice would be any good. They always conclude, "No. Ball or spherical dice would tend to keep rolling so as to delay the game, and when they stopped the top might be between numbers. If a number was nearly straight up, but not quite, arguments would result." Logical thinking, as far as it goes.

But this thinking has been influenced by the traditions of how dice have been made in the past, with insufficient thought given to other

Spherical dice work because the regular shape required was placed inside.

ways they might be designed. The dice shown in the photograph are spherical, and they work fine!

You are unable to play with them, so I need to tell you what I found when I examined them. They rattle. They are hollow and each one has a weight inside, probably a small steel bearing ball. Now you can see that if the inner cavity has six depressions, the ball will fall into one of these holes and the die will stop rapidly with one number on top.

Now note that the inventor of these clever spherical dice recognized the same requirements for useful dice as my classes did. The classes didn't get beyond another tradition concerning dice: they assumed (without being conscious of it) that the required regular shape was going to be on the outside of the die, as it always had been. The spherical-dice inventor hurdled that mental barrier and put the regular shape on the inside. This is a beautiful example of creative thinking.

THE CALF PATH

One day through the primeval wood
a calf walked home as good calves should;
but made a trail all bent askew,
a crooked trail as all calves do.
Since then three hundred years have fled,
and I infer the calf is dead.

But still he left behind his trail
and thereby hangs my moral tale.
The trail was taken up next day
by a lone dog that passed that way;
and then a wise bellwether sheep
pursued the trail o'er vale and steep
and drew the flock behind him, too,
as good bellwethers always do.
And from that day, o'er hill and glade,
through those old woods a path was made.
And many men wound in and out,
and dodged and turned and bent about,
and uttered words of righteous wrath
because 'twas such a crooked path;
but still they followed—do not laugh—
the first migrations of that calf,
and through this winding wood-way stalked
because he wobbled when he walked.
The forest path became a lane
that bent and turned and turned again;
this crooked lane became a road,
where many a poor horse with his load
toiled on beneath the burning sun,
and traveled some three miles in one.
And thus a century and a half
they trod the footsteps of that calf.
The years passed on in swiftness fleet,
the road became a village street;
and this, before men were aware,
a city's crowded thoroughfare.

And soon the central street was this
of a renowned metropolis;
and men two centuries and a half
trod in the footsteps of that calf.
Each day a hundred thousand rout
followed this zigzag calf about
and o'er his crooked journey went
the traffic of a continent.
A hundred thousand men were led
by one calf near three centuries dead.
They followed still his crooked way,
and lost one hundred years a day;
for such reverence is lent
to well-established precedent.
A moral lesson this might teach
were I ordained and called to preach;
for men are prone to go it blind
along the calf-path of the mind,
and work away from sun to sun
to do what other men have done.
They follow in the beaten track
and out and in, and forth and back,
and still their devious course pursue
to keep the path that others do.
They keep the path a sacred groove,
along which they're compelled to move;
but how the wise old gods must laugh
who saw that first primeval calf.
Ah, many things this tale might teach—
But I am not ordained to preach.

—*Sam Walter Foss*, 1895

12

On Coining Words

This chapter isn't about hardware inventions, but it is about an innovation that is needed, and it illustrates the opposition to change that plagues inventors.

Progress toward the equality of women was mentioned in Chapter 1. We have seen helpful and also ludicrous proposals to change words that might be interpreted as sexist (a ludicrous one: we must not call women "women," we should call them "wopersons").

When I address a letter to several persons of unknown gender I always write "Dear People" or "Salutations" instead of "Dear Sirs." But there is another problem that has been a constant frustration to me.

There is a deficiency in our language that makes it very clumsy to write without reference to sex. It is not a matter of the wrong words, but of lack of certain neuter-gender personal pronouns. We can say "person" instead of "man or woman," but there are no single words to use in place of "he or she," "him or her," and "his or hers."

Many others are also conscious of this deficiency in the English language. On August 7, 1986, after writing "he or she," Judith Martin added "(we need neuter pronouns)" in her syndicated column "Miss Manners." James J. Kilpatrick, in his syndicated column "The Writer's Art," wrote on February 15, 1987, "There is no satisfactory way out of these syntactical swamps. To speak of 'his or her' and 'he or she' is to mess up a sentence. My feeble advice, when you fall into this bog, is to get out the best way you can."

Kilpatrick is right, there is no good answer, unless we *invent* one. When there is a hardware need, we solve the problem with an invention. Words can be invented too—it is called "coining" words.

"Ms." was coined some time ago because many women objected to the fact that "Miss" and "Mrs." referred to their marital status, while all men were simply called "Mr." Now "Ms." is commonplace, and it effectively serves its purpose.

NHE, NER, AND NIS

Therefore, I propose the following three simple neuter-gender words to further promote sexual equality: *nhe*, to be used in place of "he or she"; *ner*, to be used in place of "him or her"; and *nis*, to be used in place of "his or hers" and the possessive "her."

Some example sentences:

A child and nis dog were seen in the distance.

Give ner a camera for Christmas.

The university seeks a new president. Nhe is expected to be chosen from the faculty. The budget problem is awaiting ner.

Each prisoner is responsible for nis actions.

After the senator is elected, ner spouse and nhe will move to Washington.

There is room for one more passenger and nis luggage.

The meaning of "Nhe gave it to ner because it was nis," is: "A specific person of unspecified gender gave it to another person of unspecified gender because it belongs to that person." It only took nine words to say it using my new words, but I used twenty words to say it without them.

The initial letter, *n*, in nhe, ner, and nis, stands for "neuter." Beyond that there is little new to learn, and the use of these new pronouns would rapidly become automatic. I typed those example sentences without even thinking about it. The transition is made easy because the sound of nhe is like he or she, ner is like her, and nis is like his, and they are used in the same order.

One person told me that we do not need any neuter pronouns, because we could call a person of unknown gender "it." This distasteful suggestion is, to me, an example of the great opposition that many people feel toward innovation, in this case new words. As we will discuss later, this fear of the new (neophobia) and reluctance to change are human traits that tend to limit the market for inventions.

New or coined words are entering the language all the time, but none of them make the grade unless they have appeal or fill a need (sometimes only that of a particular group). If we feel the need for nhe, ner, and nis, and start using them, they will rapidly become part of our language, and one more slight against women will be a thing of the past. Fortunately, we don't have a government agency telling us which words we can use (yet). We are free to coin new words. Like inventions, if they are useful enough they will be used.

I hope dictionaries will soon show nhe, ner, and nis: third-person-singular pronouns of neuter gender for the nominative, objective, and possessive cases. But dictionaries don't coin words—they only report on the words we use. It is up to us.

I was strongly tempted to write this whole book using nhe, ner, and nis. The fact that I didn't illustrates an important point. I decided not to take the risk. I know that the introduction of any innovation or invention, whether it is good or not, will meet with opposition from a percentage of the population.

The use of these new words is a good idea whose time is long overdue, but some of you readers are going to be opposed to it. If I used nhe, ner, and nis throughout the book, I would lose some of you. I would rather keep you in the hope you will agree with some other things I will be proposing, rather than lose you in an effort to push these new words.

Any reader who feels this is all upsetting garbage need not worry. Nhe will not have to read more of it. I will not continue to bug ner with it and take up more of nis time.

13

A Puzzling Approach to Inventing

One of the ways we can improve our creative ability is by working certain types of creative puzzles. Practice in working puzzles will make us better puzzle solvers, but that in itself is not important to us, because inventing is a rather different intellectual game than recreational puzzle solving.

It is important to observe the barriers to creative thinking that prevent us from solving certain puzzles, because for the most part the barriers to trivial puzzle solving are the same as the barriers that stifle our practical creative thinking while we are inventing. Please keep that point in mind as we examine some selected creative puzzles.

If this were a class, I might assign these puzzles as homework. You are invited to try them before you look at the answers. However, the important points are not the answers themselves but how you solve them, why the problems are difficult, and what they teach you about barriers to creative thinking.

THE TREE-PLANTING PUZZLE

A farmer has ten trees that he wishes to plant in five straight rows with four trees in each row. How might he do this?

In a puzzle where we are given numbers, it is natural to try to apply some arithmetic in solving it. The problem is then one of setting up the equation correctly. In this case an impatient person might note that we want five rows with four trees in each row, and four times five is twenty trees, "but we have only ten trees, so the problem is impossible." That would be true under certain ground rules, but is it necessarily true under the ground rules as stated?

More creative and less defeatist persons will note instead that ten trees divided by five rows is two trees per row, but they need four per row, so each tree must be in two rows. The arithmetic is reversed and the conclusion is quite different. They have by their arithmetic established a necessary ground rule.

We next observe that if all the trees are to be shared by more than one row, the rows must all intersect and therefore cannot be parallel. Note that the person whose arithmetic said the problem was impossible was assuming that the rows were parallel (but he or she was probably not aware of that assumption).

Assumptions are fine if we recognize we are making them, but they are often an insurmountable barrier to success when we don't realize we made any assumptions.

To continue with the solution of our tree-planting problem, we may at this point recall that a pentagon has five lines and we need five rows. But it is immediately evident that a pentagon alone won't solve the problem, because it has only five intersections at which to plant trees, and we need ten.

Now we may suddenly and creatively realize that if we take a pentagon and extend all the sides out in both directions, they will intersect again. We now have a five-pointed star, which has ten intersections. When we plant a tree at each intersection we will plant ten trees in five straight rows with four trees in each row.

Perhaps, in some minds, the pentagon step would be absent, and the puzzle solver would think of the star shape directly. It would partly depend on which figure the solver was most familiar with, a pentagon or a five-pointed star.

We are not going to leave this little exercise in creative planting yet, because it has more to teach us.

One of my students took this puzzle home to his wife. She worked on it for a while and drew a nonsymmetrical arrangement of five straight but intersecting lines, which seemed to her to solve the problem.

American
Tree
Farm

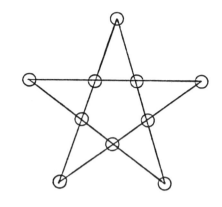

The
Lady
Was
Right

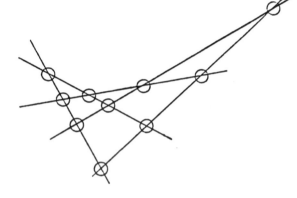

Israeli
Tree
Farm

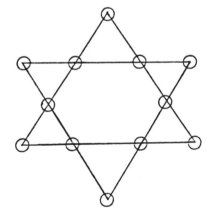

The tree-planting puzzle

Her husband, however, said, "No, that is wrong. The teacher said the answer is a five-pointed star." Regrettably, the student had forgotten that the teacher also said that most creative problems have more than one valid answer. The woman was right. She had found another solution to the problem.

My student not only forgot part of his lesson, but he showed an unfortunate blind faith in "authority," which would greatly handicap him in the field of inventing. He didn't even examine his wife's solution to see if it was correct. He didn't think he needed to, because he "knew" his expert "knew" it was wrong.

Her solution was actually an easier one than the star version, but being asymmetrical it wouldn't look as good in the farmer's orchard. The rows can be in any arrangement as long as they are straight and intersect. All we have to do is draw five straight nonparallel lines and plant trees at the resulting ten intersections, and we have correctly solved the problem. Following those simple rules it is impossible to fail.

Why, then, do most people who succeed draw a star instead of randomly placed lines? Apparently because we humans thirst for order. We frequently do things a hard way and ignore easier approaches.

Let's take this tree-planting exercise still further. Assume we had twelve trees and wanted to plant them in six straight rows with four trees in each row. Is this problem easier or harder than the ten-tree problem? I think it depends on our heritage or ethnic background.

For most Americans I think the ten-tree problem would be easiest, because we have five-pointed stars in our flag and are familiar with that geometric design. For a citizen of Israel, however, the twelve-tree problem is apt to be easiest, because Israel has a six-pointed star in its flag. For Jewish Americans the two versions of the puzzle may be equal in difficulty, because they are used to both the American star and the Star of David.

When I was having an overhead-projector chart made on the six-pointed-star problem for use in my classes, an incident occurred that illustrates another aspect of creative thinking. I had given my secretary a rough draft for her to type. It included a sketch of a six-pointed star. When she brought the finished sheet into my office, I was surprised to note that she had drawn a five-pointed star instead of the six-pointer I had sketched.

When I asked her why, she was surprised, since she didn't realize that my sketch had six points. This person was very cooperative, hard working, pleasant, intelligent, and a rapid and accurate typist, but she showed herself to be unobservant. Therefore, I would rate her low in creativity, because observation, visualization, and creativity go hand in hand.

When I mentioned pentagons earlier I was reminded of a joke that shows a kind of inattention to detail that would never be acceptable in an inventor. The small boy and his father were standing at the top of the Washington monument when the boy asked, "Where is the Pentagon, Dad?" The father replied, "See that hexagonal building over there? That is the Pentagon."

THE WINDOW PUZZLE

In one wall of his house a man has a window that is two feet high, two feet wide, and square. He doesn't like it because it doesn't let in enough light to suit him. So he removes the window and puts in a new window in the same location in the same wall. The new window is also two feet high, two feet wide, and square, but it lets in twice as much light as the old window. How is this possible?

We need to be reminded at this point that most creative problems have more than one possible answer. This puzzle is no exception. There is one surprising answer I am hoping you will spot, however, so let me dispense with some lesser possibilities so you won't be led

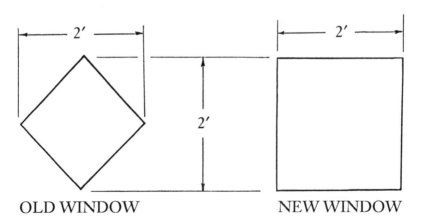

OLD WINDOW   NEW WINDOW

Window puzzle sketch

astray by them. Both the new and the original window were of plain clear glass. No trees in the yard were cut down, and no blinds or curtains were removed. Both tests were conducted on successive days at high noon with cloudless skies. You get the idea—the change was in the nature of the window.

Don't feel bad if this one stumps you temporarily. I wouldn't have given it to you if it were easy. I promise that you will do better on such problems in the future, and that is the whole objective.

The sketch shows the original window and the better window. Note that both windows are two feet high, two feet wide, and square. A little geometry will show that the second window has just twice the area of the first, so it will let in twice as much light.

I suspect that at this point many of you are unhappy with me, feeling that I have deceived you. I would point out that the only one who deceived you is yourself! You may be questioning whether the given answer satisfies the problem as stated. You will find that it does . . . exactly.

Those of you who failed to see this answer probably failed because you made two assumptions that were not in the statement of the problem, and worse, you didn't realize you made any assumptions.

When I wrote that the first window is square, you probably assumed that it was square with the wall. It is not—it is on the diagonal, but it is square.

The second assumption that you probably made without realizing it was that the original window was a normal window, and you somehow had to invent a super window that would be twice as good. In fact, the original window was weird, and all you had to do to solve the problem was to substitute a normal window.

Lesson: Beware of making unrecognized assumptions. Take the statement of any problem strictly literally. Try various assumptions, as necessary, to solve the problem, but label them as assumptions and discard them if they don't work. It is impossible to discard something you don't know you have.

THE CHESS-GAMES PUZZLE

The avoidance of unrecognized assumptions is so important to inventors that we will look at another puzzle, one that is only puzzling if we make a blind assumption: How could two men play five games of chess and each win the same number of games without any ties?

Since you had been warned and have learned your lesson, you will correctly answer that the two chess players were not playing each other. See how easy it is?

THE BOOKWORM PUZZLE

There are four volumes of Mark Twain's works sitting on a library shelf in proper order, that is, with volume one on the left. Each volume is two inches thick. A bookworm starts at the front cover of volume one and eats its way straight through to the back cover of volume four. What minimum distance did the head of the bookworm have to travel?

The obvious answer is two times four, or eight inches, but this isn't the correct answer. The correct answer is two times two, or four inches.

This turns out to be a problem in observation and visualization, both essential tools of the successful inventor. Please visualize the four books sitting on the shelf. Note that you are looking at the binding edge of the books. Where is the front cover of volume one as you see it, on the right or the left? It is on the right, next to volume two, right? And where is the back cover of volume four? On the left, next to volume three.

So our worm doesn't eat through volumes one and four at all, it only has to bore through the middle two volumes. Two times two is four. Observe and visualize wherever you can. Inventors should not take things for granted.

THE CAKE-CUTTING PUZZLE

A busy housewife has worked hard all day and has one small chore left before she can sit down and rest. She has to take a cake out of the oven and cut it into eight equal pieces. Since she is pooped, she would like to make those eight pieces with a minimum number of knife strokes. What is that minimum number?

I have given this problem to several thousand students in my classes and lectures on inventing in the past fifteen years. It is an interesting fact that the correct answer is almost always given first by a man. Why? Are men smarter than women? No way! The reason why this puzzle is best solved by men is that the puzzle is biased against women.

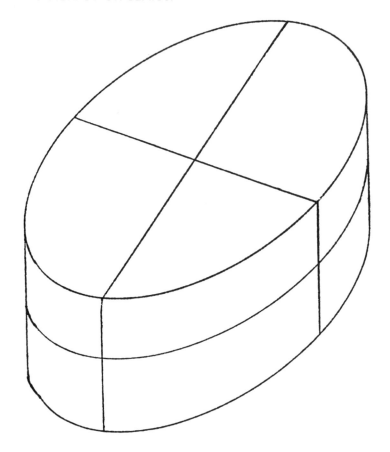

It takes only three strokes to cut a cake into eight pieces.

There are all kinds of biases in the world, which put some people or groups at an advantage or a disadvantage in some respect. Biases can and do influence creative thinking. We saw a cultural or ethnic bias in connection with the tree-planting puzzle, with regard to five-pointed versus six-pointed stars.

The most common and most serious bias affecting the aptitude of individuals is inadequate education or experience. That bias or barrier is seen when we don't know enough to solve a particular problem. But our cake problem has a gender bias.

The minimum number of knife strokes needed to divide a cake into eight pieces is three, all at right angles to each other. Visualize that and you should be able to mentally count the pieces (or look at the sketch if you must).

The gender bias here stems from the fact that women, at least in the past, have done most of the baking and cutting. You will note from the sketch that one of the three necessary cuts divides the cake into an upper and a lower layer. No one who has cut a lot of cakes is apt to even consider cutting a cake horizontally in that manner. We don't cut cakes that way. It just isn't done.

Things that are counter to our traditions or habit patterns tend to be automatically excluded from our creative thinking process. If men, at some future time, turn out to be the chief cake cutters of our society, the bias will reverse.

Such biases sometimes make outsiders more effective inventors in a field than the experts in that field. The expert knows all the things that can't be done, or that shouldn't be done, or that won't work. Usually the expert is right, but occasionally he or she is wrong. Then the biases work against the expert. The outsider lacks the negative biases, but also lacks valuable experience and knowledge in the field. Knowledgeable outsiders have the best chances.

THE CIGAR PUZZLE

Joe has the disgusting habit of picking up old cigar butts, threading them on a toothpick, and smoking them. (From a technical standpoint, such a reassembled cigar would not draw, but let's overlook that point.) Joe has found that he needs five cigar butts to make his standard budget cigar. On a particular day he finds twenty-five cigar butts, and had none left over from earlier. How many cigars can he assemble and smoke that day?

The obvious answer to this conundrum is five cigars. If the obvious answer were correct, however, I wouldn't have given you the problem. He can assemble five cigars from the twenty-five butts and smoke them. Then he will have five butts left over, from which he can assemble a sixth cigar. Look for the unexpected. It may help keep you from looking for cigar butts.

THE TWO-HANDLED TOOL ANALYSIS

I have a silly-looking tool that consists of a steel shaft with a screwdriver-type handle rigidly attached to each end of it. What is this tool used for?

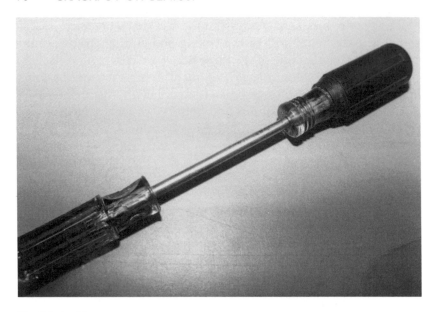

Yes, this tool has a purpose.

One might say, "This is a device ideally suited for applications requiring an apparatus of this nature." Seriously, it was not made as a joke, and it has a very useful application.

The tool is used to sell tools. As you can see from the photo, one of the handles is bare plastic, but the other has a soft rubber cover. When someone grabs the two handles in two hands and twists hard in opposite directions, the bare handle tends to slip in the hand, and the grooves in it are very uncomfortable. Yet the rubber-covered handle is comfortable and does not slip.

All a salesman of screwdrivers with rubber-covered handles has to do to convince a prospective customer is to let the customer conduct this test for himself or herself.

Most people have difficulty in arriving at the use for this strange tool by just looking at it. When I give it to a member of one of my classes and invite the student to play with it and report his or her observations to the class, however, the use of the device is immediately seen.

The moral is, if hardware is involved in a problem, play with the hardware if at all possible. By "play" I mean experiment, examine, test, feel, bend, heft, bounce, smell, listen, taste, twist, throw, or do to it whatever is appropriate to learn all you can about it.

Inventing involves hardware. An "invention" is not patentable unless it is hardware of some form. Inventors must understand the physical world. The more they play with things, the more they will learn about the way things are made.

The ancient Greek word for "play" was *paidia*, while the word for "education" was *paideia*, only one letter different. The clever Greeks had observed the close relationship between playing and learning.

Inventors, more than people of most occupations, need to play, to play with hardware as well as with ideas. Playing is often a process of trying different things, and that is creative thinking. That is inventing.

A child plays because it is fun, because he or she enjoys it. The inventor also does the playing he or she calls inventing because it is enjoyable. The inventor may also invent in the hope of making money. A very few will make money, but the returns on time invested that all inventors can count on are challenge, excitement, satisfaction, and pleasure.

THE NUMBER PUZZLE

0 2 3 6 7 1 9 4 5 8

In what order are the above numbers arranged?

We again need to remember that creative problems usually have more than one valid answer. There is one surprising and interesting answer that we are seeking here. There are also a few good answers that we are not looking for: the numbers are arranged horizontally; they are arranged with spaces between them; they are arranged in a single-file straight line; and they could be arranged randomly.

The desired answer is that they are arranged in reverse alphabetic order, from *E* for eight on the right to *Z* for zero on the left. Probably most of you who didn't spend much time on it failed to see the alphabetic order.

Why is this problem tough? Because we were given figures, yet the answer depends on the fact that those figures also have English word equivalents. Had the problem been written "zero two three," etc., most of you would have seen the answer rapidly.

To make the mental switch from figures to words, without any prompting, requires a significant creative insight. But this is what creative thinking is all about. Creative effort and resulting inventions are not easy. If they were I wouldn't be writing this book, and you wouldn't be reading it.

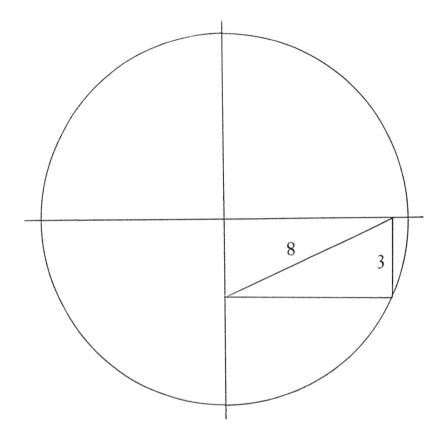

How much is the radius of the circle?

THE CIRCLE RADIUS PUZZLE

Given the accompanying figure, how much is the radius of the circle?

I don't care for some math problems, probably because math was never my strongest subject. This little problem I like, however, because it teaches us a good lesson.

You may, at this point, be trying to remember the geometry or trigonometry you once took. What formula you were supposed to learn is needed to solve this problem? Wrong approach! With any and all problems, first look for easy or obvious solutions.

We are given a circle with some straight lines inside of it. Note that several of the lines form a rectangle. Another line, which is a diagonal of this rectangle, is eight units long. All rectangles have two diagonals that are of equal length. The diagonal not shown is therefore also eight, and of special interest, is also a radius of the circle.

Since the radius is what we seek, the answer is eight, and we got it by inspection and common sense. Many inventions fall into this category. They are not difficult if we are observant and use common sense.

14

Serendipity

Serendipity: the making of pleasant discoveries by accident. The word was coined by the eighteenth-century writer Horace Walpole after he heard the ancient fairy tale "The Three Princes of Serendip," in which the main characters had frequent accidents but the accidents always worked out to their advantage.

Serendipity has played a key role in a great many discoveries and inventions. It is therefore important to study the phenomenon and learn how to use it effectively. First, a few examples, starting with a personal one.

ON JOB HUNTING

Years ago my wife was looking in the classified ads for a better position. She saw one in her field that looked interesting and tried to call the phone number. At the last second her eye lost the column she was looking at and she ended up dialing the number from a help-wanted ad in the next column over, in a different classification.

When the person answering started talking about an entirely different type of work, she did not say, "Sorry, wrong number," and hang up. She was curious, listened, thought about it, and ended up taking the wrong-number job in a field that was new to her and paid more than the job she had intended to inquire about. She did well in the new position and was very happy with it.

A somewhat similar fortunate accident happened to me only yesterday. Our mail-delivery person made a mistake and put a magazine that did not belong to us in our mailbox, a magazine I was not familiar with. Before putting it back in the box for correct delivery I glanced at it. The only advertisements in it were for popular scientific books. I called the magazine and they expressed an interest in this book. It is possible that the mail-person's accident will result in a good marketing outlet for the book you are reading. My curiosity and action was just as essential as the accident, however.

GOODYEAR

Originally, the latex from certain tropical trees was largely a curiosity. People did find they could smear it on a coat for waterproofing. This worked at moderate temperatures, but in warm weather it got sticky and smelly, and in cold weather it got brittle and cracked. Children found that if they wadded dried latex into a ball it was a useful toy, but again only at moderate temperatures. Also, someone discovered that a lump of it could be used to rub out pencil marks, so it was given the name "rubber."

In the 1830s, Charles Goodyear searched for a process to make India rubber more useful. His efforts were in vain until one day he spilled a small amount of a rubber mixture he was working with on a hot stove. Instead of ignoring the accident, he carefully examined and tested the sample that was subjected to high temperature and found it to have all the properties he was looking for. With the help of serendipity, he had invented vulcanized rubber.

The important thing to note here is that in order to profit by such accidents one must be alert to them, curious, and diligent in learning all that can be learned from the accident or mistake.

Luck almost always plays a part, however. In this case the stove had to be hot enough, Goodyear had to leave the spilled portion there long enough for it to cure, and the mixture he was working with that day had to be one that was suitable for vulcanization.

CORNING GLASS

More heat than intended also resulted in another invention. Dr. Donald Stookey, working in the Corning Glass Works laboratory in 1948, was trying to process some special glass at three hundred degrees Celsius. "Fortunately," the thermostatic control on the furnace failed and the temperature went up to nine hundred degrees. Dr. Stookey, like all good experimentalists, had the sense to examine the remains of his ruined experiment. Then serendipity struck again; he accidentally dropped it on the concrete floor. It didn't break! Further tests and analysis showed that the high temperature had converted the glass into a new ceramiclike material with properties superior to glass.

PENICILLIN

In 1928, Alexander Fleming, an English bacteriologist, was conducting studies on a particular bacteria. When he went home one night he neglected to cover one of his cultures. Upon returning the next morning and seeing his omission, he didn't simply throw the ruined experiment out—he examined it under a microscope out of curiosity. To his surprise he didn't find other contaminating bacteria from the air as he expected. Instead he found that the original bacteria in his culture had strangely been killed. Through patient further experimentation he isolated penicillium mold spores in the air, and derived penicillin, which we still use to kill bacteria.

THE EDISON EFFECT

Thomas Edison accidentally noticed a strange phenomenon in one of his experimental incandescent lamps that had nothing to do with the lamp's purpose of giving light. Edison didn't understand what he saw, but being a good experimentalist, he carefully recorded his observations. From Edison's record, John Fleming in England first used this "Edison effect" to invent the diode vacuum tube, and a year later Lee De Forest in the United States used it to invent the triode, or amplifying vacuum tube.

OTHER ACCIDENTS

Luigi Galvani, a professor of anatomy in Italy, accidentally discovered some important electrical facts while dissecting a frog in the 1780s. Louis Daguerre, a French artist, invented the original or daguerreotype photographic process in 1837 by accident. Wilhelm Roentgen discovered X-rays, in Germany in 1895, by accident. The DuPont company invented Teflon by accident. The slinky toy was found by accident. The list is long.

Sometimes the person having the accident isn't the one to profit by it. James Watt invented a better steam engine as the result of analyzing the effects of an accidental leak in one of Thomas Newcomen's steam engines, which Watt had been hired to repair.

LONG SHOTS

Sometimes good fortune comes more by chance than by accident, but only when accompanied by observation, curiosity, and hard work. As a critical part of one of my inventions I needed a material that would expand and contract in a certain way with changes in moisture.

Over a period of several years of searching I found, by test, many materials that would partly meet my requirements, but none that were good enough in all respects. One day I started reading all of the ads in the yellow pages of the phone book for any clue as to a better material for my purpose.

It was a long shot, but inventors are used to long shots. I had only gotten to the *C*s when the entry "Chemical Grouting Company" caught my eye. I had no idea what chemical grouting was, so I phoned the company and asked them. I soon learned that they use a complex organic chemical in their business that had exactly the properties I required, yet their use of the material was entirely different from my proposed use. The long shot was a bull's-eye!

15

The Raw Materials of Invention

It takes materials to build things, and this applies to building ideas as well as physical things. Ideas or concepts are sometimes said to "come out of the blue." If "the blue" is nothingness, the saying is false. You can't get anything out of an empty box.

Ideas are almost always based on simpler ideas, knowledge, and observation. The idea of a wheel probably came (many times) from observing the rolling of something round such as rocks or snowballs. The idea of a lever may have come from observing the action of a log lying across another log.

Inventions usually combine many earlier concepts. The manual typewriter, for instance, used the ideas of the wheel, the lever, shaft, bearing, spring, ratchet, gear, screw, ribbon, ink, keyboard, and others, not to mention the inventions of language, the alphabet, and paper. The invention of the typewriter was not possible until these previous inventions had been made and were known to the inventor.

Invention requires knowledge as well as creativity. Creativity is really the ability to assemble knowledge in new ways. The successful inventor is usually a person who observes and remembers, and therefore has a great store of useful knowledge about physical things. In other words, he or she has a head full of raw materials from which to make inventions.

To the beginning inventor, the message of this chapter is to be curious and observant, and to try to understand and remember all of the potentially useful ideas that you see, hear, or read. In most cases you won't immediately create an invention based on your latest observation, but you must collect the pieces. The invention may have to wait until after you have learned other necessary facts. Learning one new word doesn't allow you to finish writing a book, unless that was the only word that was missing.

Curious people will automatically observe and remember things that interest them. It is very important for an inventor to have a personal store of applicable knowledge, so he or she should consciously try to enhance it. The following examples illustrate my point.

INVENTION OF THE THERMOSTAT

In the early 1900s, John Spencer, a fifteen-year-old boy, had a job as boiler fireman on a steam donkey engine used in a logging camp. John observed that the door of the firebox would snap outward with

Serendipity led John Spencer to invent his Klixon thermostat.

a pop when it reached a certain temperature, and snap or "oil can" back in when it cooled a bit. He realized that the diaphragm action of the door was indicating temperature. He found that when the door popped on the cooling half-cycle it was time to put more wood on the fire.

Eighteen years later Spencer invented the bimetallic disk thermostat, which snaps to a concave shape when it gets hot, just as the old firebox door did. He patented the disk thermostat in 1923, registered the "Klixon" thermostat trademark, and received sixty improvement patents on thermostats.

Texas Instruments Company, which later bought out the Spencer Thermostat Company, still uses the Klixon trademark and still manufactures Spencer's invention, which is used, among other places, on fractional horsepower motors as protection against overheating.

The average person might have been annoyed by the snapping firebox door, if they had noticed it at all. Spencer observed it, was curious, remembered it, used it, improved on it, profited by it, and met a need with it.

THE BROKEN STICK

Years ago, sitting in a dull management meeting, I was idly flexing a wooden dowel that was used as a blackboard pointer. I flexed it a little too far and it broke in the middle. The reason for the failure was immediately evident: the grain of the wood was badly diagonal at that point.

I might have put the pieces of the stick down then, but the meeting was still boring so I subconsciously continued to play with them. In my idle playing I lined up the two broken ends of the sticks and was rolling them between my fingers.

Suddenly I observed that as they rubbed against each other there was a shearing action between the broken elliptical-shaped faces. This was an action that I have never seen in nature or in any machine, and it excited me.

So I have this idea for a rotary shear stored in my mind for possible use someday. I will probably never use it. We use very little of our total knowledge, but since we don't know which knowledge we may need, we should try to remember all of it. Fortunately, the things we remember best are the things that interest us most, and

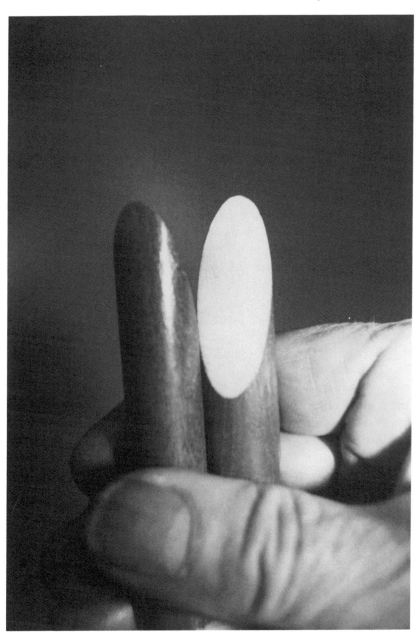

Rotary shearing action discovered by serendipity.

inventors are usually interested in physical things that may be useful to them.

My rotary shear idea could perhaps be used in a new type of food mixer or blender, to stir paint, to cut fiberglass roving into short lengths for use in spraying up fiberglass boats, or who knows what. If you can use the idea, be my guest. I have given it away but I still have it.

A FLUORESCENT IDEA

One day I replaced a fluorescent light tube in my shop. As I was putting in the new tube I noticed a small blemish on its surface, a clear spot where there was no fluorescent powder on the inside of the glass. The spot looked dark because I was looking through it into the dark interior of the tube.

But I was suddenly curious: What would the bare spot look like when I turned on the tube? I carefully noted where the spot was so I would be able to find it when the tube was lighted, but as it turned out this precaution wasn't required at all. When the tube was lit, the bare spot was very much brighter than the rest of the tube.

Since I was looking through the clear spot and seeing the inside of the coating on the far side of the tube, I realized that the inside was much brighter than the outside, a fact I didn't know and had never considered before.

A little lesson in the operation of fluorescent lights is needed here if I am to keep from losing some of you. The glass tube is evacuated except for a little bit of mercury vapor. The electric current through the mercury vapor causes it to glow mercury-vapor blue, but most of its radiant energy is in the form of ultraviolet light. To turn this ultraviolet into white light we coat the inside of the tube with a mixture of mineral powders that fluoresce when exposed to ultraviolet. It is called fluorescent light because the source of the light is fluorescence.

I suspect many of you inventive types now have the same revelation that came to me at that point. If the inside of the tube is much brighter, we could get much more light from fluorescent tubes if we coated only the top half with fluorescent powder. Obviously, the inside of the coating fluoresces brighter because the inside is directly exposed to the exciting ultraviolet.

So it appeared I had invented a much more efficient fluorescent light, but experience told me it couldn't be that simple. If it was, hundreds or thousands of bright people would have realized the same thing long ago, and the product would be on the market.

A trip to a technical library to research fluorescent lights filled in the missing pieces. I learned that some fluorescent tubes are made with the coating only on one half to get more light, just as I reinvented it! These special tubes are used in reproduction equipment and other applications where people are not constantly exposed to the light.

Again some of you are ahead of me. You see that with half the tube bare there would be strong ultraviolet radiation, which is dangerous to human health. Some people still go to tanning parlors to get skin cancer, but others even try to avoid excessive direct sunlight.

Half-coated fluorescent tubes could be sold to those who want an eighty-year-old skin on a forty-year-old body, for use in their homes in place of conventional lights. This would reduce their electric bill because the lights would be more efficient, save the cost of the tanning-parlor sessions, save the gasoline required to get there, and save time, but logic doesn't seem to be a significant consideration in such matters. More and more we are becoming a nation that believes that if it costs money it is good, but if it is free it is worthless, or at least not the "in" thing.

Having climbed down from that soapbox, I call your attention to the essential part that serendipity played in all three of the previous examples: If the firebox door hadn't snapped back and forth, Spencer probably would not have invented his thermostat. If the blackboard pointer hadn't been cross-grained and I hadn't played with it, I would not have the concept of the rotary shear. If the fluorescent tube hadn't had a flaw that I noticed and was curious about, I wouldn't know nearly as much about fluorescent lights. Observe the unusual, try to understand it, and remember it.

SLOTTED CUP HANDLES

Have you observed the cup that screws onto the top of a Thermos-brand vacuum bottle? You should have, because there is something unusual about the handle. If you don't remember it, go have a look. I will wait.

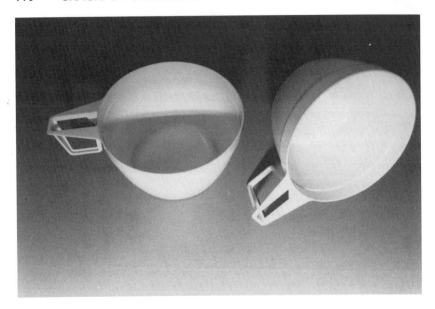

The slot in the top of these cup handles was a clever and profitable invention.

The handle has the usual lateral hole through it, but it also had a hole or slot down from the top. Why? Not only Thermos cups, but many other plastic cups have this strange handle design. The reason for the slot in the top of the handle is not that someone thought it would look good. That top hole very greatly reduced the cost of manufacturing the cup!

To understand this surprising fact we need to understand the process by which a Thermos cup is made. It is manufactured by plastic injection molding. The metal mold is made in two pieces, an outer or female half that controls the shape of the outside of the cup, and an inner or male half made to shape the inside.

The two halves of the mold are mounted in an injection molding machine and clamped together. Then hot molten plastic is injected into the cup cavity between the mold halves under high pressure. The metal mold takes the heat out of the plastic, causing it to rapidly freeze into a solid cup. The machine then separates the mold halves and the finished cup is removed.

But what about the slot in the top of the handle? Patience, I'm getting to it soon. To make the lateral hole in the handle in the usual manner, part of the machine would have to move in a direction at right angles to the main mold-opening motion, in order to pull out the mold plug that formed the handle hole.

Probably millions of ordinary plastic cups have been molded in that way, but the extra mechanism required to pull the handle-hole plug out sideways is expensive and sometimes troublesome.

By putting the slot in the top of the handle, the mold can be designed so the plug that forms the lateral hole in the handle can be pulled out from the top, in the same direction as the main mold halves move! This greatly simplifies the mold and reduces its cost, thereby reducing the cost of the manufactured cup.

It is not enough to know how a proposed invention is to work and what it needs to look like. The processes used to manufacture a product and the materials from which it is made largely determine its cost.

Since the cost is a major factor in the success or failure of an invention in the marketplace, the inventor needs to know as much as possible about all of the processes and materials that might be used in manufacturing his or her invention, and to design it for minimum cost.

NEW MATERIALS

We have been talking about ideas and observations as materials from which to make inventions. Now let's talk about concrete materials for the same purpose. Anytime a new material becomes available that has properties not available before, it is likely that inventions will crop up that use those properties. One might say that the invention has been waiting for a material to be developed that would make the invention possible.

Historically, the steam engine couldn't be invented until metals were available, electrical inventions weren't made until wire was available, books weren't invented until after paper, and so on. Not only did the necessary material have to exist first, but inventors had to know about it, have access to it, and know how to work with it.

There are several relatively new and little-known materials now available, with some unusual properties, which will be increasingly used in inventions. You might be the inventor of one of those inventions after reading here about the material you needed. You're welcome.

GORE-TEX

The first unusual material is a plastic fiber matrix developed by the W. L. Gore Company. It is sold under their trademark "Gore-Tex" but is somewhat different from their Gore-Tex fabrics used for water-repellent clothing.

This special Gore-Tex, which is sold as a joint sealant among other uses, has the amazing property of variable volume. A rubber band will stretch, but when it stretches it becomes proportionally smaller in cross section (it grows thin). This type of Gore-Tex will also stretch or compress, but it does not grow thin or fat. In other words, its volume may be changed! When it is in its expanded state it floats in water, but when it is compressed it sinks—therefore, it has controllable buoyancy as well!

The structure of the material consists of millions of microscopic fibers inherently bonded together. When it is fully expanded the fibers are extended to their full length, and the material has the tensile strength of the combined fibers. When it is compressed the fibers bear against each other, much like felt. Then it has the compression strength of the combined fibers.

So far I am not aware of any inventions that are based on these variable-volume and buoyancy properties of this special Gore-Tex, but I predict they will come.

One interesting use of the material now is for making artificial blood vessels. This Gore-Tex is made from Teflon, which is nonbiodegradable and nontoxic, so it can be used in human beings. A very important feature of this Gore-Tex for use in blood vessels is the microporosity that results from its fibrous structure. As in Gore-Tex clothing, the pores are so fine that water (or blood) can't get through. In clothing the pores do let water vapor escape from inside. In Gore-Tex blood vessels, human tissue grows into the pores so the synthetic addition becomes an integral part of the body and does not continue to depend on the surgeon's sutures for fluid-tight and strong joints.

MEMORY METALS

Another type of material that has resulted in inventions, and will result in many more, are the "memory metals." Nitinol is the best-known memory metal. The name is an acronym for "Nickel Titanium, Naval Ordnance Laboratories." The first four letters tell what it is made of, and the last three tell where it was developed.

A memory metal is one that can be bent, stretched, or otherwise distorted, but which remembers its original shape and attempts to return to that original shape when heated. The first of these metals was invented in 1958, but they are still relatively little used. Some interesting inventions depend on the unique properties of memory metals.

As an example, the solar-panel-array power supplies for satellites must be folded up in order to rocket the satellite into orbit, but once the satellite is in space, the panels must unfold in order to receive maximum sunlight.

To do this by conventional means requires hinges, springs and latches or electric actuators, and a command signal or automatic sensor system. It isn't a simple problem.

One way in which it might be made a simple problem would be to substitute strips of Nitinol for the hinges between sections of the solar panels. When the panels are folded for the rocket boost, the Nitinol "hinges" would be bent. When these strips felt the heat of the sun in space they would straighten out to their original shape, unfolding the solar array. It would require no hinges, springs, latches, actuators, sensors, or commands.

We once considered this simple system at Boeing. I don't know whether it is in use now or not—it does have a few problems.

NITINOL ENGINES

The photo shows a toy heat engine that depends on Nitinol for its operation. It consists of a plastic frame, one plastic pulley, one brass pulley, and a Nitinol-wire belt around the pulleys. When the brass pulley is dipped into hot water or heated in any other way, the engine runs, quite mysteriously.

The explanation is as follows: The memory-metal belt is originally straight, but it is bent when it wraps around the small brass pulley. Since this pulley is hot, it heats the Nitinol belt where it contacts it. Once that portion of the belt is heated, it tries to straighten out to its original shape. It succeeds at the point where it leaves the brass pulley, and in doing so applies a torque to the pulley, resulting in rotation and power.

This little motor or engine is surprisingly efficient. A larger size has been sold to cool attics in the summer. The Nitinol engine drives

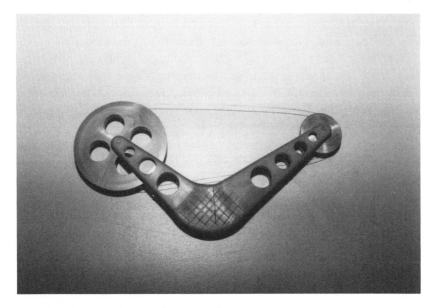

Nitinol engines may have an important future.

a fan to cool the attic, and the heat of the attic provides the energy to drive the engine.

That is still small stuff, but we may see very large Nitinol engines one of these years. A technical study has been made on the use of Nitinol engines in multimegawatt power plants! As we saw in the chapter on physics, the efficiency of heat engines is not very good. In a fossil-fuel or nuclear power plant we throw out a lot of heat because the steam turbines can't use it at the lower temperatures. Nitinol engines can.

Consideration is being given to passing the still-hot exhaust from huge power plants through Nitinol engines to extract more power. We would get many more megawatts, essentially free. This would greatly reduce fuel consumption, helping to conserve our fossil-fuel supplies and our uranium.

There could also be an environmental advantage, at least in some locations. The easiest way to get rid of exhaust heat from a power plant is to dump it into a river. But this raises the temperature of the water and tends to kill the fish. Nitinol engines would reduce the temperature of the exhaust to nearly river temperature, and the fish would be happy.

16

Inventing Doesn't Take Common Sense

Well, actually inventing does take common sense, but it takes more than common sense. To be patentable, at least, an invention has to be unique. "Unique" is the exact opposite of "common," so how could common sense alone possibly produce an invention?

In my opinion, an inventor must have a good share of uncommon sense. Let me share several examples.

THE PLUGGED PIPE

In a portland cement plant there was a large pipe that carried a mixture of clay and water (an engineer would call it a slurry). After a while the pipe plugged up due to settling out of the clay.

An engineer was called in on the problem and his common-sense solution was to put in a larger pipe. The larger pipe still plugged up, so they used still more common sense and put in a still larger pipe. If anything, it plugged up faster than before.

Then a person with uncommon sense became involved in the problem. She said, "Let's take the big pipe out and put in a pipe half the size of the original pipe." That stupid (like a fox) suggestion took a little explaining, but they agreed to try it.

The pipe never plugged up again. Some of you readers with uncommon sense see that the clay mixture would have to flow through the smaller pipe at a higher velocity, and the greater speed would keep the clay stirred up so it didn't have a chance to settle out. Our common-sense engineer would say the flow was changed from laminar to turbulent; but if he is so smart why didn't *he* think of using the smaller pipe?

THE ORE CHUTE

A smelter had a big steel chute that carried metal ore down to the smelting furnace. The bottom of the steel chute kept wearing out because the ore was mostly rock, and it was very hard and abrasive.

A common-sense person said they should use harder steel. They tried that, but it didn't extend the life of the trough bottom much. Enter the person with uncommon sense. His novel recommendation was not to make the chute bottom harder, but to make it softer—the opposite from what common sense dictated.

They lined the chute with soft resilient rubber, which made the ore roll and bounce down instead of slide. The chute never wore out again. We can see a close parallel to this solution in the design of tires. Old horse-drawn wagon-wheel "tires" were just bare steel rims. How long would they last on a concrete highway at sixty mph? Resilient rubber tires last forever by comparison.

THE ESCAPEMENT WHEEL

This last example involves a personal experience. At Boeing many years ago I managed an electro-mechanical design group. One of our projects involved a little escapement wheel, something like those in a grandfather's clock, but for a high-speed mechanism. The design we came up with is the one on the left in the photo. The wheel and teeth are molded from nylon plastic. In use the wheel rotates at a rather high speed, then a pawl or stop snaps in, catching a particular tooth, and stops the wheel.

It worked fine for a while, then teeth started stripping off the wheel. The impact load on the teeth, due to the sudden stopping, was too great for the relatively weak plastic. The common-sense answer to this problem would have been to make the escapement wheel out of something stronger, like heat-treated steel. We consid-

By using uncommon sense, this wheel was made stronger by making it weaker.

ered that, but we decided on an uncommon solution instead. We made the wheel weaker instead of stronger.

The wheel on the right is the improved model. It is still molded out of nylon, but we filled it full of holes. Now when the pawl drops in and stops a tooth, the rim stops almost instantaneously, but the hub rotates a small fraction of a revolution farther and stops gradually.

The spokes, which we ended up with as a result of the row of holes, all bend slightly during stopping, absorbing the impact load. In other words, we built in shock absorbers (not by adding them, but by taking something away).

This solution had another advantage over going to a steel wheel: the plastic wheel was even lighter than the original one, so it had less inertia. The steel wheel would have been much heavier and therefore harder to stop. We never lost another tooth.

17

Perseverance

You have all heard that Edison is supposed to have said, "Invention is 1 percent inspiration and 99 percent perspiration." Perspiration, persistence, and perseverance, the three *P*s. I once believed that perseverance was the most important trait of a successful inventor, but I changed my mind.

BAD PERSEVERANCE

Certainly perseverance is very important, but let's look at the other side first. A high percentage of would-be inventors are too persevering. They persist in working on ideas that are worthless. These people are ignorant of the laws of nature and/or the laws of the marketplace, or they are blinded by their own enthusiasm.

Many "inventors," for instance, persist in trying to invent perpetual-motion machines. We talked about that in more detail in Chapter 6, but to sum up here, perpetual motion is impossible. Those who persevere in its pursuit are wasting their time and money, and frequently the money of other people.

This is shaky ground, however. Inventors do and should challenge the experts, and occasionally they win. In the opinion of the experts, we will never see a perpetual-motion machine that works, but the experts don't know all there is to know. It is remotely possible that a source of energy may be found that is something we do not now understand, which will seem like perpetual motion. In the meantime, my money is on the experts.

The other area of excessive perseverance, stemming from ignorance of the market for inventions, is far more common. Again, we will address this problem in detail in other chapters. But, as a preview, almost all inventors are extremely fond of their latest idea, to the degree that their good judgment is affected, even if they are aware of the hard facts of commerce.

One fact is that the bulk of "new" ideas are worth a dime a dozen. There are an awful lot of very creative people out there, and there is therefore no shortage of good ideas.

Another problem is duplication of good ideas. In most cases dozens, hundreds, or thousands of widely separated people will think of the same good idea. Yet each one of those many people is apt to believe that he or she is the only one to have thought of it.

I have been discussing good ideas, even though most ideas for inventions are not good, all factors considered. Some of them won't work, some of them would cost too much to make, some of them would result in dangerous equipment, some of them too closely duplicate things already on the market, some of them offer little or no advantage, some of them would adversely affect the environment, and most of them would be of far less interest and use to the consumer than perceived by the inventor. Only one idea for an invention in two or three thousand is successful in the marketplace.

Therefore, the most important initial job of an inventor with a new idea is to try to prove to himself or herself that this idea would not pay off, so effort can be devoted to other ideas that may have a chance. The weeding-out process is of prime importance if we are to avoid wasting excessive time and money.

On this subject, I made "Quote of the Month" in the March 1987 issue of *Inc.* magazine, by saying in a lecture, "When the horse is dead, get off." (Actually, I had read it someplace.) We must recognize that most of our "horses" are going to die before their time. The problem is to separate the probable losers from the possible winners, and to use euthanasia on the losers. Feeding useless horses is an extravagance most of us can't afford. Don't persevere in keeping losing ideas—persevere in getting rid of them.

GOOD PERSEVERANCE

Thomas Edison was not a highly educated or a particularly brilliant man, but he was a great inventor. One of the reasons for his greatness was his perseverance. While he was working on the incandescent lamp he is said to have tried six thousand different carbonized plant fibers in searching for the optimum carbon filament.

When Edison was working on the nickel/iron storage battery (the forerunner of our present nickel/cadmium battery), he conducted 10,296 experiments before success! The story was told that at one point a friend was talking with Edison and expressed sympathy that the storage battery invention was not going well. Edison replied, "What do you mean, not going well? I've found 6,460 things that won't work!"

On the face of it, that remark seems silly, but let's examine it. Assume there was a total of twelve thousand things that could have been tried toward solving the problem. Let's also assume that one or more of those twelve thousand attempts would work. Edison had already tried and rejected over half of them. He had made tremendous progress.

When the inventor doesn't completely understand what he or she is working on, which is frequently the case, the work is apt to proceed by trial and error. Then perseverance in continuing to try is crucial to success. In many or most cases, success is never attained, but those are the odds in the inventing game.

Frequently the test model doesn't work and has to be redesigned and rebuilt and retested; and redesigned and rebuilt and retested; and . . . you get the idea. On one of my inventions I had to redesign and rebuild the hardware seven times before it would work. (That, by the way, is the same success rate as the 7-Up inventor experienced. Obviously 1-Up through 6-Up were unsuccessful.) That invention of mine was eventually very successful. (So was 7-Up.) The perseverance paid off, in both cases.

On another one of my inventions, I built not seven, but thirty-two prototypes. Many of them worked a little bit and were satisfactory in some respects, but they weren't adequate in other respects. I worked on this invention from October 1971 to November 1983 (and I have nine full invention notebooks to prove it), but I was never able to completely solve some of the technical problems.

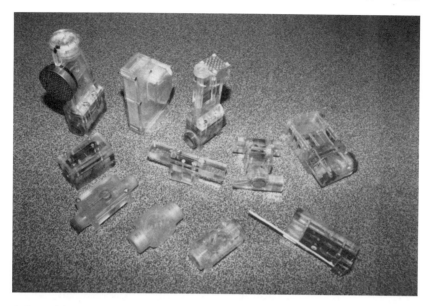

A few of the developmental models of one of the author's inventions.
Clear plastic permits study during operation.

It took me those twelve years of perseverance to find out whether I could make this one into a successful invention. At the later date I concluded that further perseverance would not solve the problems but only cost more time and money. This invention was a loser (like the vast majority of inventions). A lot of perseverance is often required to separate the winners from the losers.

MURPHY'S LAWS

These laws, more formally known as statements confirming the inherent perversity of inanimate objects, are of course the reasons why perseverance is required in inventing. Because of their impact on the inventor, the following laws are noted. This is by no means a complete list of Murphy's Laws; new ones are still being discovered.

Murphy's Laws (Abridged)

1. Anything that can go wrong will.
2. Of the things that can't go wrong, some can.

3. The thing that will go wrong is the one that will do the most damage.

4. No matter what goes wrong, there is someone who knew it would.

5. If there is a way to do it wrong, we will.

6. Nature always finds the weak spots.

7. The bread always lands butter-side down.

8. Things always go from bad to worse.

9. Efforts to prevent the above will make things worse faster.

10. If everything seems to be going well, look again.

11. If the prototype works perfectly, the production units will fail.

12. It is impossible to make things foolproof—fools are too ingenious.

Learning of these immutable facts, intelligent persons will give up trying to invent, and therefore stop reading this. But in the event that any of you insist on persevering, I will continue writing.

Seriously, it is of interest to note where Murphy's Laws came from. Captain Edsel Murphy (honest) formulated the first of these laws in 1949, at Edwards Air Force Base. Everything was always going wrong there, so he had plenty of examples. In fact, the military had recognized the phenomenon before that. The acronym SNAFU ("Situation Normal, All Fouled Up") originated during World War II.

And then there is Hogan's Corollary to Murphy's Laws: "Murphy was an optimist!"

USE AND MISUSE

The world is full of careless, reckless, impatient, and stupid people. When they get their hands on it they will abuse your invention in every way possible and a few ways that are not possible. Therefore, it is not enough to design and build a product so that it can be used right—it must be designed so that it cannot be used wrong!

Murphy says that you can never outsmart the fools, because they are too ingenious, but you must try. Persevere in trying to eliminate all of the ways that anyone might misuse your product. Try to think like a fool to see what abuses you might dream up.

18

Words of Wisdom for Inventors

Success requires the intelligent application of failure.
> —M. L. Anderson, transatlantic balloon pilot

Good judgment comes from experience. Experience comes from bad judgment.

Make your mistakes in test tubes, make your profits in vats.
> —A. C. Gilbert, maker of chemistry and Erector sets

Measure twice, cut once.

A fellow was asked, "Which is worse, ignorance or apathy?" He replied, "I don't know and I don't care." (He would never invent.)

The galleries are full of critics. They play no ball. They fight no fights. They make no mistakes because they attempt nothing. Down in the arena are the doers. They make many mistakes because they attempt many things.
> —M. W. Larmour

There are no rules around here! We are trying to accomplish something!
> —Thomas Edison

Most new ideas fail. One out of ten might be worth actually trying. And of these, one in ten might lead to something important.
> —Luis Alvarez, Nobel physicist

A mistake proves that someone stopped talking long enough to do something.   —*Phoenix Flame*

Cut-and-try is no virtue when the problem could be solved more easily by analyzing it; conversely, trial-and-error is no vice when the alternative is empty speculation.   —J. Gregory Krol

I have learned to use the word *impossible* with the greatest caution.   —Wernher von Braun

Virtually nothing comes out right the first time. Failures, *repeated* failures, are sign-posts on the road to achievement. The only time you don't fail is the last time you try something (and it works). One fails forward toward success.   —Charles Kettering

An expert is a person who, through his or her own painful experience, has found out all the mistakes which can be committed in a given field.   —Niels Bohr, atomic physicist

You will never stumble onto something while sitting down.
—Charles Kettering

Where there is no risk there is no achievement.

Don't improve it until it doesn't work.

Ability is God's gift to man; achievement is man's gift to God.

BASIC VS. IMPROVEMENT INVENTIONS

If you want to get into the history books, invent something basic. The inventors we remember invented completely new things like the airplane, telegraph, and telephone. These things did not exist at all before, but now we couldn't get along without them.

Coming up with a useful basic invention is quite difficult these days, however. The simple things have mostly been done already. In 1899, Charles H. Duell, U.S. commissioner of patents, advised President McKinley to abolish the U.S. Patent Office because "everything that can be invented has been invented."

Obviously Mr. Duell was premature, but a high percentage of the *basic* inventions had been made by 1899. Roughly five million United States patents have been issued since Charles Duell wanted to close up shop, but a very high percentage of these are on improvement inventions. For every basic invention there will usually be hundreds or thousands of improvement patents. I wonder how many hundred thousand improvement patents there are on the automobile.

It is much easier to improve on something than it is to create that something out of nothing to start with. And improvement inventions are of great importance. Most of our basic inventions were quite worthless in their initial form. The first flight of the Wright brothers was shorter than the wingspan of a 747. Bell's first telephone could barely be heard in the next room.

The improvements on the basic inventions are responsible for making them into mass-produced essential parts of our civilization. I take issue with the saying that "necessity is the mother of invention." None of the basic inventions were necessary when they were invented, because they were next to useless initially. In my view necessity is the daughter of invention.

Most of the inventors of our great basic inventions died poor. It takes so long to improve a basic invention to the point of real usefulness, get it mass-produced, and develop the market, that the patent has expired and the inventor is apt to be dead before the invention is profitable. If the original inventor makes money it will usually be on improvement patents he or she has gotten on the invention, or on the manufacturing of it.

So, my words of wisdom to the beginning inventor are: Invent something basic if you want to get famous, but invent an improvement on something if you want to get rich. If the basic invention on which you improve is already in mass production and the world couldn't get along without it, and your improvement would make it distinctly better, you are in an excellent position to sell or license your patent rights to a company that manufactures the basic product.

19

Inventors Need Workshops

An invention is not an idea—it is a physical thing.

As we shall see, a bare idea cannot be patented. It is the details of the hardware built around creative ideas that are patentable. Therefore, inventors need workshops.

Some would-be inventors have lots of good ideas but lack the talent, interest, and/or facilities to physically develop and build prototypes of their "inventions." Such people are at a disadvantage.

I put the word *inventions* in quotes because until there is designed hardware there are no inventions, only ideas for inventions. People with only an idea are not inventors in the complete sense of the word.

Also, to be frank about it, "Ideas are worth a dime a dozen." It is the experimenting and developing that takes the time, money, and effort. There is no shortage of creative thinkers; the main credit for inventions must go to the energetic, risk-taking, persevering people who develop ideas and get the hardware produced and sold. In practice, the financial rewards go largely to the doers, not to the dreamers.

Only rarely does a good idea work out when it is tried. The hard work is in testing all the possible combinations of things that might work and searching for the one or several that will work or work well enough to be competitive in the marketplace.

It is possible to get patents on inventions that have not been physically built, but in most cases it is not advisable. The hardware of the world got where it is today by constant cut-and-try refinement. The inventor who patents without building and trying the invention is taking a very big risk that the invention won't work as patented.

In my consulting business I have seen cases of inventors applying for a patent before building any hardware. Several had to revise the application or start over when later hardware efforts showed it couldn't be built, or wouldn't work, as originally disclosed.

PROFESSIONAL MODEL MAKERS

I have a small business to aid inventors and will make prototypes and do test work for them upon request. There are many others all over the country who do similar work for other inventors. *Don't hire us if you can help it!*

You would be very much better off if you could manage to do your own tinkering, testing, and model building. One learns about hardware by playing with the hardware. When the inventor is doing the dreaming and someone else is doing the hardware work, the inventor is never able to know as much about the hardware part of the invention as he or she should.

When the experimenter is someone other than the inventor, it is as though the inventor has a split personality. The thinking half is separated from the doing half. In this situation there will usually be a number of conferences between the inventor and the model builder to keep the inventor posted on what the hardware work is showing and to decide on future courses of action. But these conferences are expensive to the inventor, and at best there is incomplete communication.

This is unfortunate, because the details are the heart of the invention. The builder actually becomes more knowledgeable about the invention than the inventor, because the builder is really doing part of the inventing, the making-it-work part.

In two cases in my practice the methods proposed by the inventor did not work, and I suggested different solutions that tests showed would work. So I became a coinventor. My name is on the patents along with the original inventor's, because I made inventive contributions. To have left my name off the patents would have been illegal.

In these cases the original inventors didn't lose anything financially, since the patents were assigned (belong) to the inventors who hired me. They didn't realize when they sought my services that they needed any more inventing, but in actuality they had a coinventor for hire as well as a hardware-builder for hire, at no increase in fee.

LEARN ABOUT HARDWARE

The inventor who can do his or her own hardware development will be much more efficient, is apt to end up with a more finished invention, will get the invention developed more rapidly, and will spend less money. But to do all this, the inventor needs to play with things, learn to use tools, have a place to work, and understand materials and processes. A big order, but one that most successful inventors take for granted.

The opposite problem occurs when the would-be inventor has shop facilities and is good at making things, but doesn't have the necessary technical knowledge to understand what he or she is trying to do. In some cases such persons will hire an engineer to sort out the theory and the technical design for them, then they will do the building and tinkering.

Obviously, the complete inventor, and the one who will have the easiest time of inventing and the most chance of success, is the one who has both the necessary technical education and the shop ability and experience.

A major factor in the improved efficiency of the complete inventor is that very few drawings are required. Builders of their own designs seldom bother with formal drawings. They design as they build, and the finished hardware is the "drawing" (a three-dimensional one). Drawings will eventually be required if the invention goes into production, but only a final set. If all of the cut-and-try drawings don't have to be made, they don't cost the inventor time and money.

Another reason for doing it all yourself is the joy, satisfaction, and sense of achievement it will give you. To build something with your hands fills a human need that is seldom satisfied in modern society. Even children, today, frequently do nothing more creative with their hands than play video games. To be able to say "I built it myself" is a great ego boost.

A WORKING MODEL IS JUST THE BEGINNING

I am a believer in beginner's luck—at least things often work that way. It is frequently not difficult to make something new work the first time, even though the inventor may not yet really understand the principles behind what he or she is trying to do.

In my own case I have had good success with first tests or first models. But then I am, often for a long time, unable to repeat my success! The second model sometimes doesn't work, and I have a great deal of trouble figuring out why it doesn't work. Most of the mistakes, and therefore the learning, seems to come after a fleeting blush of success.

Also, the first model is usually tested under more-or-less ideal conditions. Out in the harsh, cruel world the product won't be as pampered. In fact, it will be abused.

ENVIRONMENTAL TESTS

Any new product needs to be able to survive and operate under all the different conditions it may be subjected to. If it might sometime be used in the arctic (or in Minnesota in the winter), it had better be designed to work at 60 degrees below zero. If it might be used in the desert, it must be able to survive sand storms as well as temperatures of 150 degrees or more. If there is a chance it will be taken into tropical jungles, it needs to resist fungus growth and not be bothered by 100 percent relative humidity. Near any seashore, or in a boat, it must not corrode from salt spray.

If the invention is to be used in vehicles, or even carried in vehicles, it must survive the vibration. If some klutz might drop one someday, it must not break. If the device is electrical, it needs to work at 100 volts and 140 volts as well as 120 volts, because our power sometimes fluctuates.

"Environmental testing" must be done to find out which of these weak spots the invention has (Murphy usually provides weaknesses in most areas); then the design and construction are improved to eliminate the weaknesses. These environmental tests are difficult and time-consuming. Count on spending a lot more money and effort in making the invention reliable, serviceable, and universally suitable than you did in making it operate for the first time.

HAVE A COMFORTABLE WORKSHOP

Comfort . . . ah, sweet comfort! But sleeping and goofing off is not the type of comfort I'm speaking of. I'm a firm believer in the value of comfort in the workplace.

Inventing is a fascinating activity for inventors, but it is sometimes frustrating, and it does take ambition and energy. In this age of the easy life there is a temptation to bag anything that requires effort and to read or watch TV instead.

If your shop is cold, damp, dark, dirty, or overly crowded, the lure of less demanding activities becomes much stronger. But if your invention-development space is a pleasant place to be, you will spend much more time in it and get to the patent office faster. I spend many hours a day in my workshop, and I love it.

First, of course, you need a space in which to work. I did a lot of invention development in my college dormitory room, and later in a studio-apartment kitchenette, but a dedicated workshop is, of course, preferred. The motivated inventor will make do with whatever facilities he or she may have and often accomplish amazing things with very little.

Which tools you will need in your workshop depends on what you want to build—but note that most inventions are made of metal or plastic rather than wood. The basic hand tools you will need are screwdrivers, pliers, a hammer, hacksaw, soldering iron, tin snips, and drill. Power tools save time, but are not essential for simple projects.

Workshops grow. None of us have anything to start with, but if one loves tools and loves to build things, the quality of his or her workshop usually improves over the years. I still have a few crude tools that I had when I was about ten years old, but as I and my

The author's invention-development workshop

family have moved from place to place, each new workshop I've acquired has been better than the one before.

20

Surveys and Searches

MARKET SURVEYS

The risk is so high in introducing any new product that it behooves us to try to find out what kind of a chance that product would have in the marketplace before we spend much time, effort, and money in developing and promoting it.

Investing in a patent is probably the largest early expenditure for most private inventors, so it is wise to have a market evaluation before committing to a patent application. Remember, only one patent in seventy ever pays for itself!

We previously pointed out the danger in trying to get a market survey of an invention at an invention fair. Asking your friends and relatives what they think of your invention would be even worse. These people have much more incentive to be supportive, loyal, kind, and agreeable than they have to be strictly objective.

Any market survey effort made by the inventor alone is apt to be of limited value and probably badly optimistic. It is very difficult for us inventors to separate ourselves from blind faith in our efforts. If we hear negative things about our inventions at all we promptly tend to discount them or forget them. The mother is biased regarding the beauty of her child.

There are professional organizations in the business of providing market surveys. Some of them are good, but the good ones are expensive and the poor ones are worthless. I don't have any easy answers for you, only words of warning.

A true story is in order at this point. During World War II all of the airplane manufacturers were building military airplanes. As the end of the war approached, these manufacturers started planning their futures as civilian airplane builders.

An organization of private airplane manufacturers (to be) originated a questionnaire that they sent to every military pilot they could find. The questionnaire asked such questions as: Do you like to fly? Do you intend to fly for pleasure when the war is over? Would you like to own your own airplane? What features and performance would you like in your airplane? What price would you be willing to pay for it?

The response was very good, and a high percentage of military pilots said they would own their own airplane. So the lightplane manufacturers started to tool up to supply airplanes for this huge postwar market.

But, when the war was over and the GIs came home they didn't get the high-paying jobs they had counted on. They got married and started having kids, and their wives said, "No way are you going to buy an airplane."

The real market was a small percentage of the market of dreams. Most of the lightplane manufacturers went bankrupt.

PATENT SEARCHES

Patent searching is looking through expired and active patents for other inventions very similar to your own.

Your patent attorney will almost always strongly advise having a professional patent search made before he or she writes a patent application for you. The search is very helpful because from it the attorney will know what other inventors have claimed on inventions similar to yours. These "prior art" patents will be cited in your patent application, making it more credible.

From this "prior art" information the attorney determines which part of your invention is really new (novel), and therefore determines what claims can be made for your invention. The patent examiner at the patent office will double-check the novelty of your invention by conducting a separate patent search when he or she processes your patent application.

So patent searches are important when getting a patent; but a preliminary search will also serve several very useful purposes at or near the beginning of the development effort on an invention.

Almost always, inventors feel they are the first ones to think of a particular idea—but almost always they will be wrong. A patent search nearly always shows that a lot of creative people had the same bright idea earlier.

If a preliminary search shows that existing patents cover all of the ideas you currently have, then you have not invented something, but reinvented it. In that case, the search served you well, since you will immediately drop further effort on the invention and save time, money, and further disappointment.

If the patent search shows that you do have some good new claimable features, you can proceed with assurance. If there are active patents (ones less than seventeen years old) that cover part of your invention, you must either bypass them by designing around them, or you may negotiate with those inventors for the right to use their basic invention in your improved invention. Also, expired prior art patents may contain some good features that you hadn't yet thought of for your invention. Since these are now "in the public domain," you can use them free with no obligation to anyone.

HOW TO CONDUCT A PATENT SEARCH

Your attorney will not personally do the patent search but will engage a professional searcher who works in the public search rooms of the U.S. Patent and Trademark Office in Washington, D.C.

Searches can be very limited and inexpensive, or thorough and costly. Except in special cases, a medium-depth search is recommended. Such a search currently runs from $150 to $700, depending in part on how much the attorney charges for obtaining and analyzing the search results.

If you do not have a patent attorney or want to save a little money, you can engage a patent searcher yourself. Look in the yellow pages of a large-city phone book under "Patent Searchers." There will usually be a number of listings, most of them with Washington, D.C., phone numbers.

I hesitate to recommend this direct route, however. In dealing with strangers, especially in another city, there is danger of a rip-off. Patent searches are frequently offered by the "invention-promotion phonies," to be discussed in Chapter 26, "Selling Your Invention."

It is safest to contract for your search through a local professional who has used and has faith in a particular Washington, D.C., searcher . . . or do the preliminary search yourself.

Most big-city libraries and the engineering libraries of large universities are apt to have a patent section. Those listed as "Patent Depository Libraries" are the most complete. Patent librarians will help you get started in making a search, but remember that it is a somewhat involved process and does take a lot of time.

THE CLASSIFICATION SYSTEM

The U.S. Patent Office has a system under which all patents are classified by subject, so they can be located. There are over four hundred classes, and these are broken down into subclasses, sometimes hundreds of subclasses per class. There are over fifty thousand classes and subclasses combined. The first step in searching an invention is to determine under which class or classes and under which subclasses the prior art might be filed. The library will have thick classification manuals for this purpose and/or a computer data base.

THE *PATENT GAZETTE*

The *U.S. Patent Gazette* is the official U.S. Patent Office newsletter. It is published every Tuesday and lists every new patent on its date of issue. (Patents are also only issued on Tuesdays.) Therefore, a complete collection of the old newsletters, back to the beginning of the patent system, will list all the patents that the United States has ever issued.

Searchers therefore look in back issues of the *Gazette* for patents in subclasses of interest. Other reference volumes tell them the dates under which to look for particular patents.

The *Gazette* publishes a summary or abstract of each new patent, listing the patent number, the name of the inventor, and what the invention is all about. This is enough to tell a searcher whether a patent is similar to the invention under development and whether it is of interest or not.

If a patent summarized in the *Gazette* needs to be studied in more detail, the full patent or a microfiche of it may be available at the library. A personal copy can sometimes be made at the library, and a personal copy can always be ordered from: The United States Patent and Trademark Office, Washington, D.C., 20231. Send the patent numbers desired, and three dollars (as of 1993) per copy for postpaid delivery.

The experience of personal patent searching is interesting and educational. But if your time is worth much, you will be ahead in ordering a professional search. Also, an amateur searcher will miss a high percentage of the prior art because of his or her inexperience. An amateur search can locate enough prior art to give the inventor a feel for what is out there, but it will by no means be as complete as a professional preliminary search or, especially, the search the patent examiner will make.

21

Invention Notebooks

All inventors should keep notebooks on their inventions. We will examine the several reasons for this and present the recommended rules for preparing these notebooks.

LEGAL DOCUMENTATION

If you apply for a patent on your invention and if, as sometimes happens, another inventor applies for a patent on an invention essentially the same as yours at about the same time, the patent office will declare an "interference" action between the two patent applications. Interferences occur with about 1 percent of all patent applications.

Your patent attorney and the other inventor's attorney will be notified of an interference, and they will be asked to present certain data to the patent office to help the patent examiner decide who is the true inventor and therefore who should get the patent.

Your attorney will be asked to document your "conception date," the "reduction-to-practice date," the filing date, and perhaps to prove that you showed "diligence" in the development of your invention. The best and most legally acceptable way he or she can do this is by the use of your inventor's notebook, if your notebook has been properly kept.

Your conception date is the date on which you first had a complete concept of your invention. Your reduction-to-practice date is the date on which you successfully demonstrated that your invention would work, either by an actual working model, by acceptable analytical methods, or by filing an application. Filing is accepted as "constructive reduction to practice." If you worked on your invention seriously and developed it as fast as you reasonably could under your individual circumstances, you showed diligence in the eyes of the patent office.

All of these facts, and many others, should be duly recorded and legally witnessed in your invention notebook. If all of the information the patent office needs to resolve the interference is there, in ink, and the competing inventor didn't do well in keeping his or her notebook or didn't get it witnessed or has no notebook, you are apt to be the winner of the patent, even if the details of the competing inventor's invention are just as good or even better than yours. If you have a good notebook and the other inventor doesn't, you may be the winner even if his or her conception date or filing date is earlier, or both!

If the other inventor thought of the idea first or made it work sooner or worked harder but can't prove any of it, he or she has a problem. Your well-kept notebook will prevail over his or her unsupportable claims. Invention notebooks are the recognized and accepted vehicle for proving these important points with regard to invention disagreements.

NOTEBOOK USE IN LITIGATION

It also sometimes happens that after you have been granted a patent, some other inventor or company accuses you of infringing on one or more patents they hold, or you may think they are infringing on your patent. The resolution of such disagreements, whether in court or out of court, will again be strongly influenced by the presence of and the quality of inventor's notebooks on one or both sides. These notebooks are customarily submitted into evidence as prime exhibits in patent litigation.

NOTEBOOK RULES

To assure that the patent office and the federal courts will accept your notebook as a true account of the facts regarding your invention, you must follow some simple but important rules in preparing the notebook:

The notebook is to be bound, not loose-leaf. It is to be a complete and factual chronological account of the development of the invention. Therefore, the pages are sacred in the sense that none may be removed and none may be added out of sequence.

The pages of the notebook must have printed numbers or be numbered in ink, and all entries must be made in ink. There is to be no erasing, deleting, or otherwise obliterating anything that has been written or sketched. You are expected to make mistakes. All inventors do. But you must not try to cover up your mistakes, because they are an important part of an invention's development. If you present a notebook that shows no errors or false starts, it will have little credibility, because the invention process is one of cut-and-try, and the courts and patent office know that.

If you write something in your notebook that you later decide is wrong, or is a poor approach, don't erase it or tear out the page. Simply go on and explain why you have decided you made a mistake and what you now propose instead.

It is sometimes desirable to glue things like photographs, special engineering curves, and advertisements onto pages of an invention notebook. Each such item must be permanently attached to the first unused page as soon as it is available. By all means, glue in photographs of all test equipment and prototype hardware you build in connection with the invention.

If the invention is a technical one, engineering curves may be needed in the development. I use notebooks in which the pages have been ruled with rectangular coordinate lines (ruled vertically as well as horizontally). These are very handy for plotting simple curves right in the notebook, but sometimes I need to plot a log-log, semilog, or polar-coordinate graph. In these cases I plot the curve on a separate commercial form and glue the curve sheet into my notebook, giving it the same page number as the page to which it is attached.

SIGNATURES AND WITNESSES

To minimize challenges as to the honesty of your notebook, you must not only sign and date it yourself, but get two or more competent friendly witnesses to read, sign, and date the pages of your notebook as you complete parts of it. Get the witnesses' signatures as soon as possible after writing anything important in the book. Proof of early date claims can be critical.

The inventor's dated signature is normally placed at the lower right-hand corner of the page, and the witnesses sign at the lower left-hand corner. It is a good idea for you to sign and date every page, but it is essential that you sign and date, and have your witnesses sign and date, in ink, all pages that show your initial conceptual thinking, further creative ideas, progress in the design, tests, hardware development, and so forth.

Right above the space you reserve for the signatures and dates of your witnesses, write, "Read in confidence and understood by:" Most authorities simply suggest "Read and understood by:"; but I feel the "in confidence" addition is useful. For one thing, it puts your witnesses on notice in writing that this is a confidential disclosure to them. For another, it may help assure the patent office or the courts that you did not publicly disclose or "publish" your invention. We will discuss that point in detail in a later chapter.

By all means have your witnesses sign and date all photographs that you have glued into the notebook, right on the face of the photograph (use matte-finish prints or take the gloss off the signing area with an ink eraser first). This will prove that the important hardware shown in the photograph existed on or before the date your witnesses signed the photo.

Don't use family members for invention notebook witnesses. Relatives are much more apt to be suspected of perjury. Pick your witnesses carefully. Try to avoid elderly people, and those in poor health. Your witnesses may be needed to testify many years later, and you want them to still be around when you need them. Likewise, choose responsible, credible, stable people.

If your invention is technical and somewhat difficult to grasp, attempt to choose witnesses that understand what you are trying to do. They may be called on to discuss technical aspects of your invention in court, so the "understood by" note above their signatures is not just words. Another side benefit may come from useful thoughts these bright witnesses may suggest when you discuss your invention with them. Three heads are better than one.

Do not get a notary public to witness your notebook. Notaries seldom read what they sign; they are only witnessing your signature. You need more than that. Even if you did ask them to read it you could not expect them to remember you or your invention or to understand it.

Also, do not send yourself a registered or certified letter. That only establishes a date, it does not give you a witness. Beginner inventors are sometimes very concerned that someone may steal their invention. Such fears are usually groundless. Your witnesses should be trusted friends. Even if they decided to steal your invention they would be in a very poor position to do so because you have their signatures in your notebook acknowledging that the invention is yours. We will discuss the stealing of inventions in more detail in Chapter 25.

YOUR USES OF YOUR NOTEBOOKS

In practice, an invention notebook is seldom used as a legal document, because patent interferences and patent litigation only occur with a very small percentage of the patents issued. If you decide to apply for a patent, your patent attorney may want to study your invention notebook to help him or her prepare the patent application in an optimum manner, but you are the one who will use your notebook the most.

In most fields of endeavor we must write records of some kind in order to keep track of what we are doing. Inventions, especially involved ones, are certainly no exception. Some brains retain more than others, but none of us have perfect memories, and the details of our inventions must not be lost. I use my invention notebooks a great deal, constantly referring back to earlier pages or earlier volumes to refresh my memory on the details of just what happened. The past determines the future.

All inventions seem more simple when we first think of them than they actually turn out to be. I'm reminded of the saying, "If we knew in advance the trouble we were going to have with most inventions, we would never have the courage to start."

It is a fact that most of the things we try in connection with most of our inventions won't work, or at least won't work well. Inventing, therefore, is frequently a process of elimination, trying and rejecting many approaches that don't pan out, searching for the way or ways that will work.

Let us look at a hypothetical example, one that is not unlikely. Suppose you tried the most obvious way of accomplishing your invention, but it didn't work. You figured out why it didn't work and tried another approach, and that didn't work either. So you tried method number three, which also failed. This went on for a dozen times or more over a period of several years. You are not discouraged, because you have the perseverance of an inventor. You review your notebook for clues on where to go from here and suddenly realize that you could make an approach work now that you tried way back in the beginning, because you have learned in the meantime how to solve the problem that stopped you then.

So you carefully read the entries on that early effort to refresh your memory on the details and rerun the test with the required changes. Neat, orderly, and efficient. On the other hand, imagine that you had kept no notebook. Your chances are poor of ever remembering that early test. Even if you did, you wouldn't remember the details, and it will cost you a lot of time and effort to regenerate them, if you can at all. Keep good notebooks; you will never regret it.

Because of the great usefulness of my invention notebooks to me, I include all kinds of material in them. Certainly all the important dates, photographs, etc., and also test setups, test data, results, marketing thoughts, information on competitive inventions or products, advertisements on parts or materials that might be useful in the invention, and just my daydreaming on the subject.

I am sometimes asked if an inventor can put more than one invention in one notebook. That is up to you. I am frugal of both paper and storage space, so I have a catch-all notebook for small inventions and to record ideas I haven't done anything with yet. On the other hand, many inventions end up taking more than one notebook volume. One of my inventions took nine volumes, covering twelve years of effort.

On each of my inventions I always end up with file folders of material, letters, etc., in addition to the invention notebooks. This is probably necessary, but we must be careful to get all material important to the invention process itself into an official notebook where it is better preserved and bears the dated signatures of the inventor and witnesses.

22

Types of Personal Monopoly Protection

Article I, Section 8 of the United States Constitution states, "The Congress shall have the power . . . to promote the progress of science and useful arts by securing for limited times to authors and inventors the exclusive right to their respective writings and discoveries."

The exclusive right given to authors is the copyright, and to inventors it is, of course, the patent, sometimes called a letters patent. An additional related type of exclusive right granted by the government is the trademark. In fact, the full name of the patent office is "The United States Patent and Trademark Office," or the PTO, a part of the U.S. Department of Commerce.

Copyrights are not handled by the patent office but are instead registered and filed by the Library of Congress. Trade secrets are not granted by the government at all, but they are recognized by the government.

Patents, trademarks, copyrights, and trade secrets are all personal property and go by the more fancy name "intellectual property." The courts recognize intellectual property, and citizens or companies can sue others to defend their intellectual property, the same as for real or other personal property. Intellectual property can be bought and sold, mortgaged, used as collateral, specified in contracts, licensed, or even stolen (as in patent infringement).

TRADEMARKS

A trademark is granted (registered) by the PTO upon request, but only after the product has been sold in interstate commerce. The individual states also register trademarks. Registration of a trademark is not essential under the law, but it is highly advisable. Certain rules apply as to what types of things are permissible in a trademark, and a computer search is made to assure that a requested mark is unique. See your patent attorney for details.

Before registering a trademark you may print the initials TM after your chosen trademark and use it, but you may lose this mark if it is later declared unacceptable or already used. Likewise, the initials SM are used after an unregistered "service mark," used to identify a service.

After a service mark or trademark is granted federal registration, the initial *R* (with a circle around it) or the words *Registered Trademark* are used after the mark. Federal trademarks and service marks have to be renewed every ten years, and they expire and can be adopted by another company if they are not used for a period of two years or more. With proper renewal and defense they can be maintained indefinitely.

There is a tendency for the trademarks of popular products to become the commonly used name for the product itself, even for products made by competitors of the trademark holders. For instance, *nylons* is the word most women use for stockings. "Nylon" was once the registered trademark of the DuPont company for their brand of polyamide plastic.

DuPont provided the polyamide plastic from which the first stockings of this type were knitted. These DuPont "Nylon" stockings wore much better and soon displaced both silk and rayon stockings. When other chemical companies started supplying other brands of polyamide filament to the hosiery mills, the ladies still called these stockings "nylons."

The trademark Nylon had gone generic, so DuPont abandoned it. The new trademark for DuPont polyamide is "Zytel." Have you bought any Zytel stockings lately?

Other former trademarks have also gone generic (become common words that do not denote a specific brand). The words *thermos* and *aspirin* were formerly trademarks of the Thermos company and of the Bayer company. Xerox, Band-Aid, Formica, Kleenex, Plexiglas, Lucite, and Styrofoam are still trademarks, as far as I know, but they have been greatly weakened by improper usage.

Two interesting examples are "Coca-Cola" and "Coke," both trademarks of the Coca-Cola company. "Coke" especially is under pressure from careless usage, but the Coca-Cola company has vigorously defended it.

At one time my daughter Pam worked at a lunch counter that served Pepsi Cola and did not have Coca-Cola. A man came to her counter and asked for a Coke. Pam poured him a Pepsi without thinking about it. He informed her that she did not serve him what he asked for and demanded to see the manager. She was told that in such cases she must tell the customer she was sorry they did not carry Coca-Cola, and then offer a substitute.

That person, and probably many others, was an employee of the Coca-Cola company. His job was to try to keep retailers from abusing Coca-Cola trademarks, in a campaign to prevent them from going generic. I heard that one retailer refused to cooperate with Coca-Cola's request. The company took the retailer to court and won.

In a 1983 U.S. Supreme Court decision, the word *monopoly*, in connection with the game, was declared generic. Briefly put, the Parker Brothers toy company lost their monopoly on Monopoly.

COPYRIGHTS

Copyrights are used to give some protection to a vast array of personal works, including all types of artwork, writing, music, and performing. If it is intellectual property, and it's not an invention, a trademark, or a trade secret, it is probably copyrightable. Ideas cannot be copyrighted. The copyright is on the specific way something is said, written, drawn, shaped, or performed. Copyrights are good for the life of the copyright holder and an additional fifty years.

As with trademarks, no official action is required in order for a company or individual to declare a piece of work copyrighted. All you have to do is write "Copyright," the year, and the name of the owner on the work and on all copies of it. A capital letter *C* with a circle around it is often used in place of the word *copyright*.

Although you don't have to, you may register your copyright by applying to the copyright office. It would usually be difficult to bring suit over claimed copyright infringement unless the copyright is registered.

A copyright registration made within three months of publication or other release of the work gives the holder certain legal advantages. See your patent attorney or regular attorney if you have questions.

To do it yourself, write to Register of Copyrights, Copyright Office, Library of Congress, Washington, D.C., 20559. You must submit a filled out application, a twenty-dollar fee, and two copies of your work. (If your work is a twenty-ton granite statue, try a couple photos of it.)

There are different forms for different types of copyrightable works. To get the right one, phone your local U.S. government information office, or, as of this date, phone the copyright office hotline, at (202) 287-9100.

A copyright, like a patent, gives quite limited protection, and then only if defended. The copyright holder must be the one to challenge supposed infringers. Literally, a copyright says that someone cannot make "a copy" of the work without authorization, but what constitutes a copy? That question is often settled in court. It doesn't take a very major change in the plot, the setting, or the names of the characters of a story to constitute a new work that does not legally infringe a previously copyrighted similar work.

The doctrine of "fair use" allows others to quote short passages from a work for purposes of review, education, and reporting. Aside from what the law allows, in my experience most copyright holders are happy to grant the right to quote from their works if proper credit for the quotation is given. I obtained several such permissions in connection with writing this book.

The American Society of Composers, Authors and Publishers (ASCAP) is an organization of, and for the benefit of, copyright holders. In the mid-1900s, ASCAP sued the national radio networks and major radio stations across the country for copyright infringement.

Specifically, ASCAP pointed out that a radio station or a network would buy a single copy of a copyrighted record and play it over the air. The copyright holder would receive a small royalty for the sale of that one record, but millions of people could hear the recording without further royalty compensation to the owner. ASCAP argued that this constituted illegal copying of the protected works and that it was unfair to the copyright holders. The courts agreed. A system was then set up for compensating the copyright holders of broadcasted works.

As the warning labels remind us, the copying of commercial videotapes is also a violation of the copyrights.

SEMICONDUCTOR CHIP PROTECTION

Integrated circuit semiconductor chip masks can be registered by the copyright office through a procedure similar to copyrighting. Ask your patent attorney about the Semiconductor Chip Protection Act of 1984.

The owners of mask-work originals may exclude others from reproducing the mask or from importing or distributing semiconductor chips that were produced using the mask. This is of great importance to semiconductor manufacturers in protecting themselves against unfair domestic and foreign competition.

Semiconductor chip protection is effective for ten years. The integrated circuit package must bear the words "mask work" (or the symbol *M* or *M* with a circle around it) and the name of the owner.

TRADE SECRETS

There are a few instances where getting a patent on an invention is not the best way to protect it or to optimize one's profit from it. If it can be determined how a new machine works or how a new product is made by looking at it or otherwise analyzing it, then a patent is the only way to protect it.

On the other hand, if the product contains important ingredients that are difficult to analyze, or if it is made by a unique and nonobvious process, a trade secret will probably be a better choice than a patent. Also, sometimes a trade secret can be used to protect something that is not patentable.

The formula or recipe for Coca-Cola is said to be a secret. That secret is over a hundred years old and still serving the Coca-Cola company well. If there was a patent on the formula for Coke, it would have expired over eighty-five years ago. Since no agency grants trade secrets, they don't expire.

A secret can be lost, however. Clever analysis by a competitor, theft, or carelessness on the part of the secret holder may destroy it. If the secret is lost by carelessness, there is no liability on any outsider's part; but if it is lost by criminal activity such as burglary, extortion, or bribery, the laws provide the same kinds of redress as for the criminal loss of other types of personal property.

DISCLOSURE DOCUMENTS

A disclosure document isn't monopoly protection by itself. It is an optional but useful preliminary step toward getting a patent. The "Disclosure Document Program" of the U.S. Patent Office is a service to inventors. A disclosure of the invention is sent to the patent office, where it is filed for a period of two years. This document is proof that the inventor had the invention at least as of the date of the document.

The disclosure document is not a substitute for a patent or a substitute for an invention notebook. Rather, it supplements the notebook and provides the inventor with an additional witness, the patent office itself.

A disclosure document may be submitted before the inventor has decided whether to apply for a patent or not. If a patent is applied for on the subject of the disclosure within two years of the submission of the document, a separate letter should be sent to the patent office connecting the previous document to the patent application. If no patent is applied for in two years, the document is purged from the files. It is possible to get a patent after that, if patent rights haven't been lost for the usual reasons, but the document cannot be used as proof of a particular conception date in a later patent.

The fee for filing a disclosure document is nominal. For the latest fee information, and for instructions in filing a disclosure document, write to the Commissioner of Patents and Trademarks, Washington, D.C., 20231.

Beware: Filing a disclosure document does not change any of the rules with regard to losing patent rights! One year after public disclosure of the invention by any of many means, or by public use of the invention, the right to get a patent on that invention expires, regardless of any disclosure document.

23

What a Patent Is

There are three basic types of patents, sometimes called "letters patents," which cover at least six different kinds of patentable inventions. New and useful asexually produced living plants are granted "plant patents." Ornamental but useful designs are granted "design patents." And inventions on processes, machines, articles of manufacture, and compositions of matter are granted the most common type of patent, "utility patents."

I will completely ignore plant patents, except to note that in recent years genetically engineered original animal inventions may also be patented in some cases.

DESIGN PATENTS

Design patents are another category of limited interest to the average inventor, so I will cover them briefly. Then we can get on to utility patents, which most of you are dying to hear about . . . I hope.

A design patent is a bit like a copyright in that it is concerned with the artistic, ornamental nature of a work, in this case a useful invention instead of a work of pure art.

Design patents, which are good for fourteen years, are very short, easy to get, and relatively inexpensive. But, like copyrights, their protection is quite limited. They protect only the external appearance of a product or invention against exact copying—things like furniture, lamps, and silverware patterns.

Design patents are not intended for things that are copyrightable. Simple inventions in which the appearance is the most important feature might be covered by a design patent only. An invention that is covered by one or more utility patents may also have a design patent if the inventor or his or her attorney feels it is advisable.

For instance, Boeing probably has a design patent on the 747 jet airliner, which will hopefully keep McDonnell Douglas from marketing an airplane that looks exactly like the 747. In addition, Boeing probably has several dozen utility patents on technical features of that same 747.

PATENTABILITY REQUIREMENTS

In addition to falling into one of the above categories, in order to be eligible for a United States patent, an invention is supposed to be novel, nonobvious, workable, useful, legal, and it must be something physical.

How new or different must something be to be "novel"? This question results in much disagreement between patent examiners and patent attorneys. The patent office requires that a new invention offer significant improvement over previous similar inventions. Small or obvious advances are not patentable. "Nonobviousness" is also a difficult thing to define exactly. Many inventions are very obvious to most people once the inventions have been disclosed; but if they were really obvious, wouldn't these inventions have been made a long time ago?

Because of the novelty requirement, an invention cannot be patented if it is known or used by others in the United States or if it is patented, published, or otherwise publicly disclosed in any country before the applicant claims to have invented it.

THE ONE-YEAR RULE

Further, if an invention has been publicly disclosed, advertised, or on sale by the inventor for more than a year before the patent was applied for, it cannot be patented in the United States. In Canada this grace period is two years, but in most other foreign countries there is *no* grace period.

So beware! If you may want foreign patents on your invention, they must be applied for *before* the invention is "published" (publicly disclosed by any means, including by issuance of a patent on the invention). Most countries (under international treaty) will accept the United States patent application date as the date of the foreign patent application if that application was made within a year of the U.S. filing. But that means that at least the U.S. application must be made before disclosure of the invention, or foreign rights are lost.

Talk to your patent attorney early enough to get foreign protection, if you want it. Foreign filing can be done only after the U.S. Patent Office grants a foreign filing license. This is usually granted when the filing receipt is provided by the patent office.

WORKABLE, USEFUL, AND LEGAL

The submission of a working model was once required as part of a patent application. The model would either prove or fail to prove that it was "workable." Now models are seldom required. If the invention seems feasible and doesn't violate laws of science, the examiner will usually not challenge its workability.

Likewise, the usefulness of an invention is seldom challenged by the patent office. If the invention could conceivably be somewhat useful to someone for something, it is officially "useful." This certainly does not imply that the patent office is saying the invention will be successful. Most inventions are not successful in the marketplace.

If the invention does or contributes to something that is illegal, it is not patentable. If you have invented a new tool for picking pockets, you will have to struggle along without patent protection on it.

IT MUST BE "HARDWARE"

If you will examine the list at the beginning of this chapter of things that can be patented, you may notice that they are all tangible or physical things. The patent office requires that an invention involve hardware in some way in order to qualify for a patent, but in their broad interpretation "hardware" includes cloth and other soft things. In other words, inventions involve "material" things. In patent language it is called "physical embodiment."

In the past, computer software was not patentable, perhaps because is was software, not hardware. Software is copyrightable, and I think it is now also patentable under certain conditions.

The patent on an invention is always written around the details of the hardware, not on the unique ideas behind the hardware. Ideas alone, unsupported by physical implementation of those ideas, are unpatentable.

Mathematical formulas, methods of doing business, and so on, are not patentable because they do not employ physical things. A scientific discovery is not patentable because the "invention," if any, was made by God or nature, not by the discoverer.

In my teaching and consulting I am sometimes asked by someone how he or she can get a patent on something they have seen on the market (frequently foreign) that they don't think has a U.S. patent. No, no, no! Only the inventor can get a patent, and then only when the invention is new.

"Patent medicines" are physical things, they may be novel and useful, and they might be thought to qualify for patents under the compositions-of-matter category, but they usually don't. "Patent medicines" are not patented! Normally they are nonpatentable simple mixtures of nonprescription ingredients, but they are often protected by trademarks and as trade secrets.

In order to qualify for a patent, a mixture of ingredients must have useful properties other than what one would expect from the properties of the ingredients alone. Metallic alloys frequently have properties different from those of their ingredients, and so they are patentable. Man-made chemical compounds always have properties much different from the elements they are composed of, and they are also patentable.

PATENTS ARE CONTRACTS

A contract is a legal agreement between two parties that gives something or some advantage to each party. A patent is a contract in which the advantage received by the inventor is seventeen years of monopoly rights to the invention. The government (actually, the people) receives a full disclosure of the invention.

Patents are not secrets—just the opposite. It is necessary to avoid public disclosure of an invention prior to applying for a patent (but see "The One-Year Rule"), but once the patent issues (has been granted), the invention is disclosed to the world.

To the world? Yes, it works that way. We are talking about a United States patent, but our patent system, like those of most countries, is open. That means that foreigners and foreign companies and countries can order and receive all the copies of our patents they want.

But what about national defense? No problem: national secrets are not patentable. The inventor of a secret military weapon may be compensated by the government, but there is no released patent.

Utility patents are in effect for seventeen years from date of issue. After that they are "in the public domain," and anyone can use them with no royalty payment or other obligation. Patents are not renewable.

Both active and expired patents are called "prior art" by the patent office. Both must be searched to determine the novelty of a potentially patentable concept.

A patent doesn't give the inventor the right to manufacture, sell, and use an invention. He or she already has that right, unless the invention infringes on someone else's patent. The right the patent conveys to the inventor is the right to exclude others from manufacturing, selling, or using the invention. The patent itself doesn't prevent other people from making or selling the invention—it only provides the inventor(s) the legal basis for taking to court any alleged infringers of the patent.

Many people believe that it is all right to make a patented invention for their own use without permission or without paying royalties. Not true—that is still patent infringement. In practice, many individuals do illegally use patented inventions and get away with it, because it is be nearly impossible for an inventor to prevent it, considering court costs, and the other complications.

Eli Whitney's invention of the cotton gin was an economic disaster for Whitney even though it was a wonderful and very much needed innovation. The problem was that it was too simple and too easy to copy. Cotton planters made their own gins with never a "thank you" to Mr. Whitney.

THE PARTS OF A PATENT

The first page or pages of a patent are the drawings, but the first written page is much like the copyright page of a book. It contains the patent number, the date of issue, the title, the name of the inventor or inventors, the application number, the filing date, the classification of the invention, a list of the most related prior-art patents, the names of the examiners, an abstract of the invention, and a note as to the number of drawings and the number of claims granted.

The descriptive title of the invention given on the patent won't necessarily be the same as the name the inventor gave the invention. Invention names are interesting. In many cases the naming of a basic invention requires the coining of a new word, for the simple reason that if the invention never existed before it never needed a name before.

In earlier times it was common to name inventions by combining words or root words, such as air+plane, tele+phone, sub+marine, etc. In recent years acronyms are more popular for invention names. LASER stands for Light Amplification of Stimulated Emission of Radiation, LED stands for Light-Emitting Diode, and RADAR stands for Radio Direction and Ranging. There is a silly story that RADAR was so top secret they spelled it backward so people wouldn't know what it was.

After the summary page in the body of the patent comes the "disclosure," which may take anywhere from a part of a page to many pages. The disclosure, the part of the patent contract that discloses the invention to the world, must be clear, complete, and tell how a "person of ordinary skill" can make and use the invention.

Beginner inventors are sometimes tempted to keep a few of their secrets by not putting them in the disclosure. That is a strict no-no! Keeping important parts of the invention out of the disclosure will result in preventing the patent attorney from getting as strong a patent as might otherwise be obtained, and it is also fraudulent and illegal.

PATENT CLAIMS

The last part of a patent is the "claims," by far the most important part as far as the inventor is concerned. The claims, and only the claims, give the inventor the monopoly rights to the invention. The disclosure is the government's side of the patent contract; the claims are the inventor's side.

The inventor (or almost always the patent attorney) will claim in the patent application the specific features of the invention that are believed to merit exclusive rights to the inventor. The patent office may grant all or some of these claims in a patent. If no claims are granted there will be no patent, since that would be a one-sided contract, one providing no protection to the inventor.

An independent claim describes the complete invention and stands alone. A dependent claim describes only a part of the invention and therefore depends upon another claim for the complete invention.

All patents must have at least one independent claim, but they may have many, and many dependent claims. There is no official limit to the number of claims permitted in a patent, but it will usually range from one to twenty-five or so.

The worth of the patent to the inventor is determined by the claims. A greater number of claims will normally mean a stronger patent, and the strength of the individual claims is equally important.

So what is a strong claim? A strong claim is one that is "broadly" written so that it would be very difficult for another inventor to invent around it. A weak claim is one that is so "narrowly" written that anyone "skilled in the art" could easily find another way of doing the job of the invention without infringing on the patent claim. An entire claim can be avoided if only one element of the claim is avoided.

All inventors and their attorneys try to get the strongest patents they can, meaning the most and strongest claims. On the other hand, the patent office, represented by the patent examiners, restrict the claims granted to no more than the novelty actually shown in the disclosure part of the patent.

Everything claimed must have an antecedent in the disclosure; therefore it is very important to make the disclosure as complete as possible. If the attorney or the inventor thinks of another claim, it can be added to an application already submitted, if it is already covered in the disclosure. If the desired additional claim does not have an antecedent in the disclosure, it is not allowed, and the disclosure is not allowed to be expanded. This situation would require an additional or new patent application.

The writing of good claims is a difficult art that takes lots of study and practice. That is one of the aspects of getting a patent where the help of a professional is highly advantageous.

In my invention classes I have the students try to write a claim for the common wire paper clip, and we discuss their attempts in class. To have a professionally written claim on the paper clip for an example, years ago I had my friend and patent agent, Rob Jenny, write one for me.

Only recently I found a claim on the paper clip written by patent attorney Thomas R. Lampe, again as an example. I was impressed by how remarkably similar the two claims are, even though they were written by two individuals with no contact with each other. Rob Jenny's claim follows:

> I claim: A device for holding at least two sheets of material together, comprising:
>
> a first element having a first planform and lying in a first plane,
>
> a second element having a second planform and lying in a second plane, and
>
> means for resiliently connecting said first element to said second element and positioning them relative to each other so that said first and second planes are essentially parallel and in close proximity to each other and said first planform at least partly overlies said second planform.

I trust you understand all that perfectly. The patent office uses a language that is quite precise. In common English one word may mean many things; in "patentese" it means only one specific thing. Patent claims must be written in patentese.

Another peculiarity to note is that the above claim is several paragraphs long but it contains only one sentence. All patent claims are limited to one sentence, but they may be of any length and hung together with commas and other punctuation marks. Silly? Sure, but that is the rule.

We mentioned broad claims and narrow claims. Let's have a look. An example of a ridiculously broad claim on the paper clip might simply claim all ways of holding sheets of material together. No patent attorney would waste time writing such a claim, but if it was submitted the patent office would never grant it because it claims far more than the simple invention of the paper clip. As written, this claim would cover not only paper clips but glue, tape, staples, nails, rivets, bolts, welding, and soldering.

On the other hand, a ridiculously narrow paper-clip claim might specify a wire of specified diameter made of a particular steel, heat-treated to a specified hardness, having bends of a specified radius, and other details.

Assuming the paper clip hadn't already been invented, the patent office would readily grant such a narrow claim, but it would be worthless to the inventor. Anyone could manufacture clips having one small change from the clip specified in the claim, and there would be no infringement. This is called "designing around the patent."

24

Getting a Patent

Getting a patent is not difficult if the requirements discussed in Chapter 23 can be met. It is the profitable use of a patent that is difficult. Since only 1 or 2 percent of all patents ever pay for themselves, don't try to patent everything you invent. Be selective. A good gambler doesn't bet on every horse.

A few affluent inventors may attempt to patent all of their inventions, like perpetual students collect college degrees. Profiting is not the object with such people; it is the acquiring of the degrees or the patents that provides their ego lift.

Corporations tend to be very selective in what they patent. They know the odds, and they have the advantage of less emotional involvement in a given invention than a private inventor is apt to have. Corporations will usually have market experts who are better able than the private inventor to judge the worth of an invention. One major company reviews about four hundred employee inventions per year and gets patents on about forty of these, or 10 percent. They will make money on only a few of the patents.

So, whether you are trying to profit by inventing, or just collecting patents, let's examine the steps in getting a patent.

PATENT ATTORNEYS AND PATENT AGENTS

First, the matter of professional help. The patent laws permit a person to apply for patents personally, without a patent attorney, but I don't recommend it and neither does the patent office for reasons explained later. The inventor who gets his or her own patent usually ends up with a very weak patent.

There are two classes of professionals qualified to help us get patents: patent attorneys and patent agents. Both must have a degree in engineering or in basic science, both have to have taken a course of study in the rules of practice of the patent office, i.e., patent law, and both must have passed the Patent Law Bar Examination.

The difference between a patent attorney and an agent is that the attorney has graduated from law school and is a member of the bar. Since the lawyer has more education, he is able to charge more. As of this writing, in the early 1990s, patent agents' fees may average forty dollars per hour and patent attorneys' fees may average one hundred dollars an hour.

In recent years I have used a patent agent almost exclusively. My patent agent gets patents for me that are just as good as an attorney could get—at about half the price. Of course, there are inexperienced or poor agents, but no more than there are inexperienced or poor attorneys.

For getting patents, attorneys and agents are equal except for cost; but attorneys can negotiate contracts for you and represent you in court, while agents can't because they are not lawyers. Few patents are sold or licensed or end up in court, so I use a less expensive agent to get them, then hire a patent attorney to represent me legally when and if I need one.

There is also a distinction between a patent attorney and a patent lawyer, at least in some states. In order to draw up contracts and to represent a client in court, an attorney must usually be admitted to the bar of the state in which he or she is currently practicing. The title "patent lawyer" indicates membership in the state bar.

THE COST OF A PATENT

As usual, that depends. The largest cost, by far, is the cost of the attorney's or agent's time. The patent office charges moderate fees for filing and issuing the patent, drawings must be made by a patent draftsman, a patent search must be paid for, and typing and other clerical help is needed. All of these things are normally paid through the attorney or agent, but some inventors prefer to do part of the work themselves or to contract for it separately.

The total bill will depend to some degree on whether the attorney or agent charges high or low, but it will be more influenced by whether the invention is simple or complex, by how much prior art must be studied, and by how many claims it is possible to write.

I have stalled as long as I can. The total cost of a patent in 1993 may be between $2,000 and $4,000 if prepared by an agent, and between $4,500 and $10,000 if prepared by an attorney.

After a patent has been in effect for four years, a maintenance fee is due to the patent office to keep the patent in effect. At eight years a larger maintenance fee is due, and at twelve years, a still larger one. If the maintenance fees are not paid, the patent expires and goes into the public domain.

The maintenance fee system has been opposed by many inventors and inventor organizations, but I personally favor it. I think government agencies such as the post office and the patent office should be self-supporting. It is not reasonable for me to ask the taxpayers in general to support my inventing hobby, even though some argue that invention should be subsidized for the public good. I don't like subsidies (not even the tobacco subsidy).

The patent maintenance fee system is good because it taxes success. It is comparable to a graduated income tax.

By requiring maintenance fees, the fees for filing and issuance can be lower. If the patent hasn't made any money in its first four years, the inventor will probably choose to let the patent expire rather than pay the fee. If it is a money-maker, and still making money after eight years, the inventor will be happy to pay that maintenance fee so he or she can continue to make money on the patent.

Another advantage of the maintenance fee system is that it releases most unused patents into the public domain for unrestricted use by manufacturers and other inventors much earlier.

AFTER FILING

A few weeks after a patent is applied for, the attorney of the inventor will get an official filing receipt from the patent office acknowledging receipt of the application, assigning it an application number, and notifying the applicant that an examiner will be assigned to process the application.

The first communication from the assigned examiner may not come until a year or more later. The individual examiners usually have a backlog of applications, and the latest one goes to the bottom of the in-basket.

PATENT EXAMINERS

The patent examiners are educated and skilled employees of the patent office. There are many of them, and they are organized into different fields of technology. For instance, if you have applied for a patent on a valve, your application will go to an examiner who is an expert on valve inventions.

If a patent examiner and a patent attorney for an inventor fail to resolve disagreements, the matter can be referred to supervisory examiners, to the U.S. commissioner of patents, to the U.S. district courts, and even appealed to the Supreme Court in rare cases.

OFFICE ACTIONS

The assigned examiner will write a letter to the patent attorney after studying the patent application. This is called the first "office action." In other words, the patent office is taking its first action on the application.

This office action will normally consist of a critique of the claims presented in the application and an acceptance or rejection of each one. It is rare that the examiner will accept or "allow" all the claims as written. Sometimes changes in the wording of a claim will be required, often a claim will be disallowed completely, and many times *all* claims will be disallowed!

This action doesn't necessarily mean that the patent will not be granted, but it puts the burden on the attorney to prove that some of the claims, or rewritten or modified claims, are justified under the existing prior art.

These office actions are written in "patentese" language and are very difficult if not impossible for a layperson to understand. In communicating with the patent office, and in writing claims, the patent agent or attorney really earns his or her pay.

After the attorney or agent responds to the first office action, the patent examiner will prepare a second and usually "FINAL" office action in which the final claims, if any, he or she will allow are specified. The claims allowed may be fewer in number and frequently more narrowly written than those in the application. Response to a patent office action made FINAL is limited. If no claims are allowed an appeal may be filed, or an application termed a "continuation in part" may be filed, adding more details and narrowing the claims in further attempts to show useful novelty over the cited prior art.

AFTER YOU GET A PATENT

. . . you will get some mail. I would like to be able to tell you it will be from people who want to license or buy your patent, but it won't be. It will be from people who want to sell you things.

You may get an advertisement for a frame in which to keep the original copy of your patent. You, the inventor, get the only original. It will be in a cover, with a government seal, and tied with a blue ribbon. The seller hopes you will want to display it in their frame.

You are also apt to get a postcard from a company that would like to sell you a list of other patents in the same subclass as yours that have issued since your patent. This could be useful to you if you are closely following the development of the product.

Another postcard may come from a firm that would like to sell you a list of companies that are engaged in the manufacture and/or sale of products in the field of your invention. This, of course, could be helpful in trying to sell your patent. The last postcard I got of this type offered me a list of three companies. By reading the *Thomas Register* in the library I had already made myself a list of fourteen such companies, so I didn't need their help.

A fourth type of mail a new patent holder may get is from companies that wish to sell the inventor advertising space to help in the promotion of the invention. This ad-to-sell-ads tried to get me interested by actually printing a drawing from my own latest patent, in advertisement format, an appealing touch. But something else struck me much more: the flier contained misleading advertising.

It loosely talked about the yearly Geneva Patent Show, the company's participation in that show, and went on to say, out of context, "One-third of patents sold at last meet." That got my attention, since my experience is that a very small percentage of all patents are ever sold. Not liking to let dishonest people get away with it, I immediately sent them a letter.

I wrote, "I am interested in that figure. What does it mean? Are you saying one-third of the patents you advertised last year were sold, one-third of the patents at the Geneva Exhibition were sold, or what? Can you supply me with a list of these successful advertisers, including addresses and phone numbers?"

I received a reply promptly. After I had pinned them down, they wrote, "The percentage of patents sold or licensed through any system, including ours, is always low. . . . It is also not generally known how difficult and expensive it is to find companies truly interested in new patent acquisition. . . . We have no magic answers." Enough said.

25

Infringement

You have received your patent, and you are about to do something with it, when you discover that XYZ company is manufacturing what looks to you like your invention. After recovering from initial panic, anger, and whatever, what do you do about it? You can't ask the police to make XYZ stop, because patent infringement is a civil matter, and the police have no jurisdiction in it.

You should probably take a copy of your patent and go visit XYZ, or write them a letter. After hearing your story and looking at your patent they could possibly say, "We are sorry, we weren't aware of your patent. We will stop manufacturing your invention, of course, and we will also pay you royalties for the units we have already sold."

They could . . . possibly, but it is highly unlikely. Much more likely they know all about your patent and their patent attorneys have studied it in detail. They have either decided that what they are making is not infringing on your patent, or, if it might be infringing, they think the profit they are making is worth the risk of a lawsuit. So their response to your complaint is more apt to be, "We are sorry, but our attorneys feel that our production is not infringing, and we choose to continue with it."

If they have neither convinced you nor scared you off, your next step will be to see a patent attorney, preferably the one who got that patent for you. The attorney will refamiliarize himself or herself with your patent claims, become familiar with the XYZ product that has you concerned, and study all the patents that XYZ has on that product. Then he or she will be in a position to express an opinion on whether XYZ is infringing or not.

You may be surprised and shocked if the attorney concludes that there is no infringement, and he or she is apt to do just that. Unfortunately for inventors, the coverage of most modern patents is very limited. The patent protection you really gained is quite small. For instance, if the XYZ patents were granted before yours, your claims were limited by the patent office to features that they did not claim and are not using.

But let's say the gods are with you, and your attorney agrees that it looks like they are infringing. Your attorney will then probably propose to write a letter to XYZ on your behalf, stating that in his or her professional opinion the XYZ product infringes on your patent number so-and-so. The letter may go on to ask that XYZ cease and desist, or negotiate a license agreement with you, or you will take them to court. Your attorney may also mention your claim to damages for their past actions.

Don't start counting your profits yet. The battle has just started, and the odds are not in your favor. Possibly XYZ will give in at this point, if their attorneys consider their patent position to be weak. But more likely, especially if they are making money producing the product, they will again decline to take any action.

Now you can either forget the matter or sue them. The sad part is that even if your position is strong, you are still apt to lose in court, and it will cost you a lot of money for lawyer fees.

"Why?" you ask beseechingly. "Because," I reply. Because of several factors. We have a legal system but, it isn't always a justice system. You probably have much more limited financial means than XYZ, therefore they can hire more and better attorneys than you can.

That is one "because." Another one is that the courts frequently disagree with the patent office. One branch of the U.S. government grants patents, and another branch of government frequently declares them invalid. This disturbing anomaly arises from the fact that the patent office is in the business of granting limited monopolies, while the courts tend to be antimonopolistic.

So, I am sorry to report, the position of a little person trying to defend his or her patent is not rosy. The best answer to that problem is get a big person to protect your patent.

26

Selling Your Invention

INVENTION-PROMOTION PHONIES

Before we talk about what to do in trying to sell your invention, we need to talk about something you should *not* do.

Do not pay an invention-promotion or invention-development company to try to sell your invention for you. They will try to bleed you of all the money they can, but very seldom will they succeed in doing anything that is necessary to your success that most of you couldn't do faster, better, and much cheaper for yourselves.

Some such companies are outright swindlers, others try to offer a useful service, but because of the very poor odds in the invention-selling business, their results will almost always be disappointing.

The companies I am talking about can be recognized by the following: They always want money up front. They never guarantee in writing to sell your invention. They always compliment you on your invention and verbally lead you to believe they can sell it for you. They have no list of successful previous clients they can refer you to.

These companies advertise widely in newspapers, in science-fiction, science, and mechanics magazines, on radio, and on TV. Some of them actually break laws, but they are hard to control. If one is forced out of business in one city, it is apt to go to another city and start back up under another name.

According to information released by the Minnesota Inventors Congress in the fall of 1991, one invention-development firm was forced to disclose the following statistics: "Total number of customers who have contracted with us for invention development services prior to the last 30 days is 4,402. Zero customers received an amount of money in excess of the amount of fees paid for our services." No successes!

In the September 11, 1991, issue of the *Wall Street Journal*, it was reported that a Federal Trade Commission (FTC) investigation against American Idea Management Corporation (and its two successor corporations, Idea Management and Patent Assistance Corporation and Technology Licensing Consultants Inc.) will result in $570,000 in partial refunds to injured customers.

FTC filings in federal court also show that Invention Submission Corporation of Pittsburgh is being investigated. Beware. You have been warned.

FINDING PROSPECTIVE COMPANIES

This chapter will apply to both outright sale of your patent (assignment) and licensing it, as will be discussed in the next chapter. You should try to sell your invention immediately after applying for a patent. If you wait until the patent issues, you stand to lose a year or two of royalties.

The first step is to go to the library and study the *Thomas Register*. This is a reference work that will be found in all libraries except the very smallest. To be witty I might say the library has to be large enough to have shelf room for the *Thomas Register*. It consists of many huge volumes.

The *Thomas Register* is put out yearly by a private company and contains an alphabetical listing of all the companies of any size in the United States, along with information on each of them such as products, size, locations, and so on. It is also cross-referenced to list all the companies that make a given product.

The *Thomas Register*, and/or other reference works, such as *Standard & Poor's* and *Dun & Bradstreet*, will permit you to make a list of the companies that make the product basic to your invention, including such helpful information as the names of their officers, their annual gross, and their credit rating.

Now prioritize your list. That is, decide which company you would most like to manufacture and sell your invention, and put it at the top.

The company at the top of your list might logically be the largest one in the business, that is, at the top of the national sales list also. The largest company might be able to make and sell more of your invention, therefore earning you the most royalties, but then again. . . .

Suppose company number one has the largest share of the business, has had it for a long time, and takes its position for granted. The second-largest company in the business, however, would really like to be number one and would work hard to get there. If it looked as though your invention would allow them to beat out their long-time adversary, company number two might be willing to give you a significantly better contract for your rights than lazy old number one would. It is a judgment call.

CONTACTING THE COMPANY

If the company is small or medium-sized, try hard to deal initially only with the president of the company. Why? Because the president will be the decision maker as to whether to buy your invention or not. You are apt to get turned down by anyone you deal with. A no answer is the end of the line. If an underling says no, you will never know whether or not the president would have said yes. If the president likes it everyone else in the company is inclined to like it also, or at least pretend to.

If you are trying to contact General Motors and your invention is a new type of spark plug, don't waste your time trying to contact the president because he won't make such small decisions. But if your invention is a new type of automobile, your best bet is the president. Good luck in trying to get his attention.

In my experience it is not difficult for an unknown person to reach the president of a medium-sized company by telephone, if the caller sounds businesslike. Some secretaries will be far more protective of their bosses than others, however. Some executives like to have their calls carefully screened, and others don't.

TO PHONE OR TO WRITE

If you write directly to the president rather than phone, your message may have a better chance of getting to his desk (sad to say, nearly all company presidents are men), but maybe not. Secretaries are also apt to screen out junk mail. It therefore behooves you to keep your letter from looking like junk.

But aside from the possible difficulty in getting the president's attention, let's discuss the pros and cons of a phone call versus a letter for the initial contact. I feel the best choice depends on the circumstances.

One factor to consider is the relative status of the inventor and the president of the company. An inexperienced inventor may feel intimidated by the thought of talking to a person in a high position. In such cases a letter would have a real advantage.

In my own case, I am old enough and experienced enough in dealing with executives that I am not afraid to make such phone calls, but I'm sure I do a much better job of selling myself or my inventions by letter. Even politicians, who must be able to "think on their feet," occasionally "put their foot in their mouth."

A letter gives the writer unlimited time to compose and perfect his or her message. Over the phone you have a live audience in real time and must think fast and not make mistakes. Also, over the phone the listener can ask questions that may be difficult to answer extemporaneously.

You will probably want to avoid disclosing very much about your invention in the first contact. In a letter you are in control. On the phone, if you decline to answer questions you may hurt your case.

I have much more self-confidence in writing a letter because I can take all the time I need to compose my message in the optimum manner. But recognize that there are also disadvantages to writing a letter. Over the phone we have to talk well. By mail we must write well, spell and punctuate correctly, and produce a neat and professional-looking letter. But in writing a letter we can get experienced help if we need it; by phone we are on our own.

Letterhead stationery is very useful in promoting the appearance of professionalism, whether from a company or an individual. Incidentally, you can call yourself a company, have stationery and business cards made, and do business as that company legally without any formalities.

There are advantages to incorporation, which we won't discuss here, but incorporation requires officers and certain legal actions. If you call yourself a company without incorporating, you must avoid the word *Incorporated* (or *Inc.*) in your company title.

Another important factor is taxes. If your company is not incorporated, you must pay all applicable business, sales, income, and other taxes in your own name. The liability is all yours.

In the first contact I usually disclose only the general nature of the invention and request a personal audience with the company to disclose the invention completely and demonstrate my working models.

I write that I will not contact any other company for a period of two weeks to give the first company time to respond in the event it develops an interest in acquiring an exclusive right to the invention.

I specify the amount of time I would like for my presentation and for questions and answers, and I point out that I would come to the company's facility at its convenience and at my own expense.

I also say I would prefer to make my presentation to the marketing executive, the purchasing manager, the chief engineer, the head of manufacturing, and anyone else the president would like to invite, in addition to himself. If I speak to the president alone, he will be under stress trying to remember everything I say so he can pass it on to his people. He will also be wondering what questions to ask that his executives will want the answers to.

The president of a company is usually a businessperson and may not be qualified to understand the technical aspects of an invention. His main question is whether it will make money for his company. To answer that question he needs the help of subordinates.

With his right-hand people at the presentation he can relax. He may spend more time watching the faces of his people and listening to their comments than he does listening to the presentation. If his chief engineer smiles or nods, the inventor scores a point in the eyes of the president. If the marketing vice president asks a question and the inventor fails to answer well, said inventor loses that round. The president only needs to keep score.

I have had a high success rate in getting an audience with companies by these methods. Success in getting hearings, that is—success in selling the inventions is much more rare.

On one invention I tried to sell, the first company I contacted did not answer my letter. When I then phoned the company's president, he said they had no interest. The next four companies on my prioritized list were interested and heard my presentation. In each case the president of the company either wrote or phoned, and we set up an appointment. This invention, like the vast majority of all patented inventions, was never sold or marketed.

DISCLOSURE AGREEMENTS

I usually say nothing in my initial letter about a confidential disclosure, and in all cases where I have tried to sell patent-applied-for inventions to companies, no agreements regarding confidentiality were ever signed on either side.

Inventors who come to me for consultation sometimes request that I sign a confidential disclosure agreement, but the subsequent disclosures are usually on unpatented ideas or inventions.

If you try to sell an invention with no patent or patent applied for, the company will usually ask you to sign a nonconfidential disclosure agreement to protect itself against possible future lawsuits. You, of course, would like it to sign a confidential disclosure agreement, to reduce the chances of its taking your idea. It is not apt to do so (either sign your form or take your idea). Potential invention sales negotiations are sometimes stalemated by the refusal of both sides to give on these points.

But there we were talking about unpatented inventions, which, as we have seen, have no protection under the law. The creator of an invention has no legal rights to exclude others from building, using, or selling his or her invention unless the patent office grants such rights in the form of a patent.

This chapter chiefly concerns the selling of inventions on which patents have been applied for or granted. Here the inventor need have little concern that his or her invention will be stolen, because the inventor's patent defines or will define his or her rights. Likewise, the company will have little concern that it will be sued for stealing the invention, because its liability is limited to what the patent actually grants. If the patent office refuses to grant a patent, the company will have no liability, because the government has said the inventor has nothing to steal.

PRESENTATIONS

After invitation comes presentation. This is your opportunity—work hard not to blow it. Unfortunately, a lot of worthless inventions are offered for sale. And a high percentage of patented inventions are still worthless. This is one of the major reasons the odds on selling an invention are so poor. But here we will assume that you have a good, patented or patent-applied-for invention to sell.

Demonstration of one or more working models of the invention is almost a must, if the invention is such that a working model or prototype is possible. Nothing will convince a company's chief engineer that the invention will work like seeing it work. Seeing roughly what the hardware will look like will also do wonders to convince the manufacturing division and the marketing people that they want to build and sell the product.

But demonstrating models involves risks. Nothing will ruin your chances to sell your invention faster than a model that breaks down during the demonstration. To thwart Murphy's Laws, the hardware must be well designed, well built, well maintained, and well operated.

Practice with the model at home for a long time before you take the show on the road. Find and fix all its weak spots. If it uses fuses, light bulbs, batteries, or any other weak links, take spares. If it uses electric power, take extension cords.

It is usually impossible to make a hand-built model look exactly like the production item will look, but the model should be as attractive and professional-looking as possible. An amateur inventor must do his or her best to avoid looking like an amateur.

All other aspects of the presentation must also be planned and polished to perfection. I don't mean that the inventor's speech should be memorized—an extemporaneous talk is always better, but the use of notes is completely acceptable.

It is very important to think about the questions the company may ask and to have the best possible answers for these questions in mind. The questions I'm talking about are the embarrassing ones. Your invention will have weak spots and disadvantages. If you don't see any, you are letting your enthusiasm for it blind you.

The leaders in the companies you will be making presentations to are bright people. They will immediately zero in on the weaknesses in your idea. Don't try to change the subject or insist there is no problem. You need to agree with the questioner, "Yes, that is a good question, one I have given a lot of thought to. It could be a problem if improperly handled, but I believe that by doing this and this, but not that, we can minimize the potential problem if not eliminate it entirely." Act knowledgeable and sincere and you are apt to be believed.

Prepare for your presentation in all other ways, too. For instance, if you are going to use audio, video, or visual aids of any kind, prepare for them so nothing falls through the cracks. I like to use an overhead projector and transparencies in my presentations. To make sure there will be a projector available, I phone the company I am going to address and ask the telephone receptionist, "Do you people use overhead projectors there?"

If the reply is, "What is an overhead projector?" I know I have a problem. I can then try to take a projector with me, or phone and see if I can rent one in that town, or change my presentation so I don't need a projector.

THE BUSINESS PART

After the presentation is finished and the questions have all been answered, the president is apt to say something like, "OK, we seem to have covered it. You guys can go back to work; and would you like to come into my office for a few minutes, Mr. Inventor?"

If this occurs it is likely that the president wants to feel you out on the financial end of things, matters not discussed in the open meeting. He may say something like "Mr. Inventor, in the event that our company should develop an interest in acquiring rights to your invention, what kind of a business arrangement do you have in mind?"

I had foreseen that possibility the first time it happened and was ready for it. I had found an appropriate sample contract (in the appendix of another book on selling inventions), retyped it including the name of the company, my name, and a few other specific details, made extra copies, and had taken them along.

When the president asked his question I handed him a copy of the suggested contract and said, "Here is about what I have in mind, but I am not locked in on anything. Your attorneys and my attorney can get together with us, and I'm sure we can come to an agreement."

Make it clear that you recognize the president's risks and problems in taking on your invention and say that you don't expect to get rich from it, but it would be nice to get a little return on your investment of time and money. If he has dealt with many other inventors he will probably find this a pleasant change from the eager types who have a very inflated view of the worth of their inventions and a poor feel for the realities of the marketplace.

WIN OR LOSE

The last time I played the above game, the companies were almost always enthusiastic at the end of my presentations. This tells me my presentation was good. In each case the president promised to let me know in a few days if they had a serious interest in the invention, and they always did let me know. Unfortunately, they always decided they were not interested.

Having failed to sell it after many tries, I abandoned that invention. The patent issued, but since it was not profitable to me, I did not pay the patent maintenance fee when it came due, and the patent expired and is now in the public domain. That patent joined the vast majority of patents in not paying for itself.

27

Invention Contracts

It is possible under certain circumstances to sell an invention where there is no patent and will never be a patent. This is very rare, however, because the inventor has no legal rights to the "invention" in such a case and is entirely dependent upon the generosity of the buyer.

I put the word *invention* in quotes, because in one sense the idea doesn't become an invention until the patent office recognizes it as unique, useful, etc., and grants it protection with a patent.

In the event the inventor does not or is unable to apply for a patent for financial or other reasons, an interested company or second party may choose to pay for and do the work of getting the patent, but the patent must be in the name of the true inventor or coinventors.

Since the sale of unpatented inventions is so rare, this chapter will address only the cases where a patent is concerned. The following will speak of "the patent," but actually, in most cases, the inventor should try to peddle his or her hoped-for patent immediately after applying for it.

This is done to effectively extend the earning period of the patent. Once the patent has been applied for, the inventor has certain real and implied protections. Since it takes roughly two years for a patent to be negotiated and issued, the inventor has around nineteen years of effective protection instead of the seventeen-year life of the patent per se. That is, the inventor has the patent-applied-for period if he or she makes use of it and has the full period if the patent issues.

Contracts negotiated before issuance of a patent will usually be contingent upon the patent issuing. If the patent office refuses to grant a patent for any reason, the contract becomes void and royalties stop. Contracts can be written in countless ways, however. It is all up to the contracting parties.

If there is no patent, the company is legally free to manufacture and sell the "invention" without obligation to anyone, unless it has contractually committed itself to the contrary.

ASSIGNMENT

There are two basically different ways to sell rights to an invention, "licensing" and "assignment" of the patent. In the first case the inventor retains ownership of the patent, and in the second he or she sells the patent outright. If the patent is sold prior to its issuance, the name of the assignee will be given in the patent itself, as well as the name of the inventor.

The benefits to the inventor can be very similar in both types of contract. I see no real advantage of one type over the other. That is an area where the inventor can accommodate the wishes of the potential licensee or buyer.

If anything, I lean toward the sale or assignment of a patent, because the inventor then has no further responsibility except to collect the royalty checks. The buyer of the patent assumes the responsibility for protecting the patent and for paying the patent maintenance fees.

LICENSING

If an inventor decides to license the patent instead, to license a single company and promise in writing not to license others is called giving an exclusive license. Alternatively, several companies can be granted nonexclusive licenses to the patent.

Some inventors may feel that if they grant a number of nonexclusive licenses they will make more money in total, even though they probably wouldn't earn nearly as much on any one nonexclusive license as they could on an exclusive. They may well be right, but I would never grant nonexclusive licenses. Money isn't the only consideration.

The catch is that a nonexclusive license holder has little or no incentive to protect the patent against infringement, while an exclusive license holder does. Nonexclusive licensees would love to see the patent they hold a license to shot down in court, because they could then produce the invention without paying additional royalties.

A company that holds an exclusive license, on the other hand, gains valuable monopoly rights from the patent. If the patent were lost, the company would have no further protection against its competitors. It therefore has financial incentive to defend the patent against all comers.

In most cases, a private inventor will not be wealthy enough to comfortably protect a patent in court—therefore he or she needs a company with an exclusive license (or an assignment) to protect it. It is said that a patent is only a license to sue. In the competitive business world a valuable patent is apt to be challenged by infringement or other actions sooner or later.

The inventor who doesn't fight back loses. Unfortunately, a high percentage of those who do fight back also lose (see Chapter 25). The process of inventing is fascinating, but the business part of the game is usually frustrating and discouraging. The attorneys profit more than the inventors in most cases, in getting the patents and sometimes in court.

COMPENSATION

It might be logical to assume that if a patent is licensed the inventor gets royalties, but if he or she sells it he or she will get a fixed amount of money. As it turns out, either type of compensation, or a combination of both, can be provided by either type of contract.

If a company is interested in licensing or buying a patent, it would probably be happy to sign a contract giving the inventor generous royalties but no direct money. I do not recommend this, however. The fly in the soup is something called "shelving," which means putting a patent on the shelf rather than using it to produce the invention.

In some cases it would be to a company's advantage to buy or get an exclusive license to a patent and then shelve it. Imagine that the company in question has the largest share of the market in the business relating to the patent, and its present profits are good. If it uses the new patent to improve its product, the company could probably increase its share of the market somewhat, but introducing any change always involves risk of failure. It would cost the company considerable time and money. It might have to hire new people. It might have to buy new equipment and/or new facilities. And the company would have to pay the inventor royalties, which it is not now paying.

The company does not want its competitors to get the patent, however, because if that happened its share of the market would probably drop. If it can tie up the patent with a royalties-only contract, it could shelve the patent to keep it away from its competition while not disturbing it comfortable status quo and not spending a dime. The company won't owe the inventor anything, ever, because since the company is not producing the invention, there will be no royalties due. Very legal . . . almost.

The "almost" stems from an item in the contract the company signed with the inventor. All well-written patent contracts will contain an antishelving clause, but contracts are broken every day. If it is to the financial advantage of the company to shelve the patent, it may do so in spite of the antishelving clause. The inventor may sue, but lots of things can happen in a lawsuit. Anyway, the company knows the little inventor is probably too poor to sue.

The way to prevent this sad turn of events is to make it financially advantageous for the company to use the patent rather than advantageous to shelve it. Put in the antishelving clause for a thorough contract, but provide an incentive to assure that the contract will be honored in this respect.

The simplest protection against shelving is an initial amount of money that the company must pay to the inventor, and/or minimum royalties that are due whether sales of the invention are made or not. In other words, make the company invest enough in the patent that it must use it the patent order to recover their investment in it.

However, it will normally be very difficult to get a patent agreement that provides high guaranteed royalties, or a very large initial payment and no royalties. There are two reasons for this. First, most inventors expect to profit by their inventions chiefly from royalties. If you show lack of interest in royalties and want it all up front, the company will surely wonder what is wrong with the invention or patent. What negative things about it do you know that you are not telling them? You are expected to share the risks associated with your innovation.

The other reason a company will resist paying excessive money up front is that you are asking for money when you have already given them a cash-flow problem. It is probably going to cost the company a lot of money to develop and market your invention. Now you want to strain its finances still further during this start-up? Be fair . . . and sensible. The company's financial well-being is in your best interest in the long run.

For these reasons, a combination of initial payment and royalties is usually best for both parties. How much? That is the $64,000 question. The amount of up-front money can vary over a wide range, depending on the strength of the patent, the size of the company, and the expected earning power of the patent for the company. It may be a thousand dollars, or it may be ten million.

Some contracts are written so the initial sum is independent of royalties, and in other cases it is an advance on the royalties.

ROYALTIES

Royalties to ask for or expect are a little easier to talk about than initial payments, since they are expressed in percentage rather than dollars. But before we talk about "what percentage?" let's talk about "percentage of what?"

Contracts could be, and sometimes have been, written to provide a certain percentage of royalty based on the profit the company makes on the invention. The percentage numbers in those cases may look very exciting, like 25 to 40 percent!

Beware: There is a loophole the company could drag you through. The problem is that its "profit" can be as little as the company wants it to be, regardless of how much money it "makes." That sounds contradictory, but it isn't.

The profit of a corporation each year is what is left over for the stockholders after all expenses are subtracted from the gross income. Those expenses can include research and development, new facilities and equipment, or anything else they choose to invest in.

If your royalty contract with a company is based on percentage of profit, it is possible for the company to make a lot of money on your invention yet cheat you out of your royalties by spending all its earnings and legally declaring no profit.

If a company tries to be completely fair with the inventor, profit is still a poor baseline, especially for the company. It is probably producing things other than the invention, making it extremely difficult and costly to determine just how much overhead, shop time, and materials to charge against the invention in order to determine the profit on the invention alone. It would be a bookkeeping nightmare.

To avoid this troublesome area, royalty contracts are usually written around a percentage of the net, or perhaps the adjusted gross, sales. The inventor should expect royalties on monies actually received by the licensee for sales or leases of the invention article, process, or whatever, but the contract should clearly and completely define these conditions.

Five percent royalties on net sales is the most common figure agreed to in patent-assignment or exclusive-license contracts. That number is frequently used in many industries, but it is far from fixed. Some low-profit industries will grant only much lower royalties, while some high-profit ventures may grant more.

DON'T BE GREEDY

A warning may be in order here: "Don't be greedy." If you are a typical inventor you probably feel your invention will have a great market, but that is very seldom the case. Few companies get rich on patents.

Inventors, in turn, therefore seldom make money on patents, but patent attorneys always do. If getting rich is the goal, we inventors are in the wrong business. But I wouldn't swap jobs with an attorney. Mine is more fun.

Buying or licensing a patent is a calculated risk for any company; therefore, your patent will seldom seem to be worth anywhere near as much to the company as you feel it should be. It will often be wise to take what you can get. The inventor who appears to have an unrealistically high opinion of the value of his or her invention is apt to be dismissed without a hearing.

Because of the frustration and low odds in trying to sell inventions, I have tried to give some good inventions away in recent years. Even that is usually difficult to impossible. I still continue to try to invent, however. I bet you will, too.

Speaking of betting, there are certain parallels between gambling and inventing. The former doesn't interest me at all, but I am as much addicted to the latter as some gamblers are to Las Vegas.

I am not an attorney and am not qualified to offer advice on contractual matters, but I feel that the above general information will be of interest to many readers. It is strongly recommended that any inventor be represented by a qualified attorney during the negotiation of any contracts.

This voice of experience will advise you to try to avoid lawsuits, however. They run on and on, you are under continuing stress, and the lawyers get richer while you get poorer. Even if you "win" you are apt to lose more than you gain, all factors considered. The inclusion of an arbitration clause in a contract is a good way to reduce the chances of a costly lawsuit.

28

Start a Company?

Since selling most inventions is difficult if not impossible, one apparent way around that problem is to manufacture the invention yourself. If you already have a major interest in a company that would be suitable for manufacturing and/or selling the invention, you are ahead of most people. Let us assume, however, that you are starting from scratch.

Whether or not you should try to start a company depends on many things, like: Is the invention a good business risk? Are you a good businessperson? Do you know the manufacturing game? Do you have experience in marketing? Do you have enough money? Could you borrow enough or interest investors? What are your chances of success?

History shows that most people who are good inventors are not good businesspeople. The traits that spell success in business tend to be rather opposite to those characteristic of good inventors. There are some famous exceptions in the history books, but they are exceptions.

If you are not a business-type, don't start a business, not by yourself at least. You will probably be miserable as well as go bankrupt. But there is a way out for you, if your heart is set on being in business. Find a partner who knows the business end and who would complement your inventive and perhaps your technical and manu-facturing skills.

Don't take the word *partner* too literally, however. I have seen partners in business who started out the best of friends end up the worst of enemies. It is my understanding that legal partnerships tend to be very risky. I think forming a corporation is much safer, but talk to a lawyer before you take any such actions.

If your "business associate" has money he or she is willing to invest in the business, so much the better. If neither you nor your partner has money, you will have to borrow or find investors—and a fact of small business is that the person with the most money in-vested will have the most control. You may own 100 percent of your invention now, but if you don't also invest a lot of money in the business you will end up owning a minority share. Money almost always counts for more than patents.

MANUFACTURING

If you or your partner has money, the invention, and sales knowl-edge, but neither of you has much knowledge or interest in the manufacturing business, you don't have a serious problem, except that of business risk. There are plenty of machine shops, plastic-molding shops, electronics shops, sheet-metal shops, and other job shops that are looking for business.

It is easy to get things manufactured . . . if you pay for them. Getting a manufacturer to deliver products to you and be paid only when and if you can sell them would be very difficult.

MARKETING
THE FINISHED PRODUCT

If selling is a weak area in your company's skills, there are other companies that may buy or take on consignment your manufactured products for sale—wholesalers, mail-order companies, and chain stores, for instance. But it is more difficult to do business at the marketing end. The competition in the marketplace is rough. With that very brief comment I will back out of this area. I am not a marketing expert.

DOUBLE JEOPARDY

My parting advice on starting a business around your invention is, don't. Remember the odds are about seventy to one against the average patent making money. American Society of Inventors statistics show that only about one patent in seven hundred makes big money. And, 65 percent of all new businesses fail in the first five years.

The combined risk of a new business based on an invention is staggering. Don't mortgage your house and spend your nest egg, please.

GOODBYE AND GOOD LUCK

It is a shame to end this book on a negative note, but that is the way it is. I would be doing you a disservice to indicate otherwise. All too many invention lecturers and writers make it sound easy to show a profit. In the vast majority of cases it most certainly is not. The market for private inventions is very poor.

Lots of people enjoy acting, painting, ballet, and modeling as hobbies or avocations, but few make a living at them compared to the number who would like to. Fewer still get rich in these fields. It is the same with inventing, another glamorous-sounding career.

So my final words are, inventing is a wonderful, educational, and worthwhile hobby. Work hard at it and enjoy it. Most of you will make no more money inventing than I have made, so the sooner you put the financial part of it on the back burner, the happier you will be. To those few of you who *will* strike it rich with one or more inventions, congratulations!

Acknowledgments

Marianne, my wife, best friend, supporter, and critic, conducted the business of our lives and made it possible for me to work and write with minimum interruption.

My mother taught me the self-confidence to try things.

My father taught me love of science, engineering, designing, and building things. Dad was the first inventor I knew. He designed, built, and used the first electric clothes drier I ever saw, the first automatic garage door, and the first automatic coal stoker.

Gretchen Kaufman, my high school chemistry teacher, and a few others, encouraged me to think. Many other teachers taught me only facts.

Greg, my son, robotics engineer, inventor, and builder of things, challenges me when I am wrong (and occasionally when I am right).

The Boeing company provided me with a satisfying forty-year engineering management career, which included opportunities to invent and gain corporate knowledge.

Genie Dickerson, friend, writer, and grammarian, straightened out my English and eliminated the surplus commas in this book.

Rob Jenny, friend, inventor, patent agent, and fellow engineer, has taught me patent practice for thirty-five years, and checked the book for errors.

Paul Weston, friend, inventor, and airplane designer and builder, read and checked the book for technical mistakes, as he does my monthly columns.

Bibliography

Adams, James L. *The Care & Feeding of Ideas*. Menlo Park, Calif.: Addison-Wesley Publishing, 1986. Good, but limited to creative thinking.

Albaum, Gerald. *The Independent Inventor: Explorations in Invention & Innovation*. Eugene: Univ. of Oregon Press, 1976. Fair.

Alexander, R. C. *I Thought of It First: The Joys & Frustrations of Inventing*. Port Angeles, Wash.: Pen Print Inc., 1980. Fair.

American Patent Law Association. *How to Protect and Benefit from Your Ideas*. Arlington, Va.: APLA, 1981. Good.

Arnold, Tom, and Frank S. Vaden III. *Invention Protection*. New York: Barnes & Noble, 1971. Obsolete.

Ashford, Fred. *The Aesthetics of Engineering Design*. London: Business Books Ltd., 1969. Special interest.

Austin, James H. *Chase, Chance & Creativity: The Lucky Art of Novelty*. New York: Columbia University Press, 1985. Good.

Boyd, T. A. *Professional Amateur: Biography of Kettering*. New York: Dutton & Co., 1957. Good historical.

Brown, Kenneth A. *Inventors at Work: Interviews with 16 Notable American Inventors*. Redmond Wash.: Tempus Books, Microsoft Press, 1988. Excellent historical.

Brown, Harrison, James Bonner, and John Weir. *The Next Hundred Years*. New York: Viking Press, 1957. Old.

Burke, James. *Connections*. Boston: Little, Brown & Co., 1978. Good philosophical and historical.

Burlingame, Roger. *Inventors Behind the Inventor*. New York: Harcourt Brace, 1947. Good historical education.

Cheney, Margaret. *Tesla: Man Out of Time*. New York: Dorset Press, 1989. Fascinating man, fair book.

Clark, Donald. *The How It Works Encyclopedia of Great Inventors & Discoveries*. London: Marshall Cavendish Books Ltd., 1978. Good educational.

Conot, Robert. *A Streak of Luck: The Life & Legend of Thomas Alva Edison*. New York: Seaview Books, 1979. Good.

Crouch, Tom. *The Bishop's Boys: A Life of Wilbur and Orville Wright*. New York: W. W. Norton, 1989. Well-written biography.

Cooper-Hewitt Museum. *American Enterprise: Nineteenth-Century Patent Models*. Washington, D.C.: Smithsonian Institution, 1984. Good photographs of historical models.

Emmet, E. R. *Brain Puzzlers' Delight*. New York: Emerson Books, Buchanan, 1978. Fun and educational.

Feldman, Anthony, and Peter Ford. *Scientists & Inventors*. New York: Facts on File Inc., 1979. Good, beautiful book.

Fuller, Edmund. *Tinkers and Genius: The Story of the Yankee Inventors*. New York: Hastings House, 1955. Good historical.

Garfield, Patricia. *Creative Dreaming*. New York: Ballantine Books, 1974. Creativity only.

Gilfillan, S. C. *The Sociology of Invention*. Cambridge: The M.I.T. Press, 1970. Good analysis.

Gleeson, Murray A. *How to Make Money from Your Patent and Invention*. Elmhurst, Ill.: DuPage Products Co., 1974. Good, but overly optimistic.

Gunn, A. V. *How to Design Better Products for Less Money*. North Hollywood, Calif.: Halls of Ivy Press, 1976. Good.

Hart, Michael H. *One Hundred: A Ranking of the Most Influential Persons in History*. New York: A & W Visual Library, 1978. Good apples and oranges lists, including great inventors.

Hartman, Susan, and Norman C. Parrish. *Inventors Source Book.* Berkeley: Inventors Resource Center Publishers, 1978. Fair.

Joenk, R. J. *Patents and Patenting, for Engineers and Scientists.* New York: IEEE, 1982. Good.

Jones, David E. H. *The Inventions of Daedalus: A Compendium of Plausible Schemes.* San Francisco: Freeman Co., 1982. Fascinating extrapolation of science.

Judson, Horace. *The Search for Solutions.* Orlando, Fla.: Holt, Rinehart & Winston, 1980. Good.

Kettering, C. F. *Radio Talks by C. F. Kettering.* Detroit: General Motors Public Relations, 1955. Educational and historical.

Kivenson, Gilbert. *The Art and Science of Inventing.* New York: Van Nostrand Reinhold, 1977. Good but old.

Lampe, Thomas R. *Idea to Marketplace: How to Turn Your Good Ideas into Moneymakers.* Los Angeles: Lowell House, 1991. Excellent.

Lewis, David, and James Greene. *Thinking Better.* New York: Rawson-Wade Publishers, 1982. Helpful.

Lightgarn, Fred. *Basic Components of Creativity.* N. Aim Publications, 1979. Creative thinking only.

Lindenberger, Paul H. *Invention Licensing and Royalty Rate Structure.* Miami, Fla.: Inventor's Manuals, 1970. Good but old.

MacCracken, Calvin D. *A Handbook for Inventors.* New York: Charles Scribner's Sons, 1984. Good.

McNair, Eric P., and James E. Schwenck. *How to Become a Successful Inventor.* Mamaroneck, N.Y.: Hastings House, 1973. Good, but much too optimistic.

McCullough, David. *The Path Between the Seas: The Creation of the Panama Canal.* New York: Simon & Schuster, 1977. Excellent history.

Meadowcroft, William H. *The Boys' Life of Edison.* New York: Harper & Brothers Publishers, 1921. Old and biased.

National Geographic Society. *Those Inventive Americans.* Washington, D.C.: National Geographic Society, 1971. Excellent history.

Norris, Kenneth. *The Inventor's Guide to Low Cost Patenting.* New York: Collier Books/Macmillan, 1985. Fairly good.

Petroski, Henry. *To Engineer Is Human: The Role of Failure in Successful Design.* New York: St. Martin's Press, 1985. Good philosophical.

Pressman, David. *Patent It Yourself!* New York: McGraw Hill, 1979. Understates the difficulty.

Raudsepp, Eugene. *Creative Growth Games*. Orlando, Fla.: Harcourt Brace, 1977. Fun and educational puzzles.

Raudsepp, Eugene. *More Creative Growth Games*. New York: Perigree Books, 1980. More of the same.

Reefman, William E. *How to Sell Your Own Invention*. North Hollywood, Calif.: Halls of Ivy Press, 1977. Understates the difficulty and the odds.

Rollo, Vera F. *Burt Rutan: Reinventing the Airplane*. Lanham, Md.: Maryland Historical Press, 1991. Biography of the non-stop-round-the-world airplane inventor/designer. Fascinating even though poorly written.

Shlesinger, Jr., B. Edward. *How to Invent: A Text for Teachers and Students*. New York: Plenum Publishing, 1978. How-to section is poor; historical patent section is good.

Smithsonian Institution. *The Smithsonian Book of Invention*. New York: Smithsonian Exposition Books, W. W. Norton, 1978. Good historical.

Udall, Gerald G. *Guide to Invention and Innovation Evaluation*. Pullman, Wash.: State University Innovation Assessment Center, 1977. Pedantic.

Von Oech, Roger. *A Whack on the Side of the Head*. New York: Warner Books, 1983. Good, creativity only.

Vos Savant, Marilyn. *Brain Building: Exercising Yourself Smarter*. New York: Bantam Books, 1990. Fun and good mental practice.

Woodbury, Robert S. *Studies in the History of Machine Tools*. Cambridge: The M.I.T. Press, 1972. Detailed history.

Woog, Adam. *Sexless Oysters and Self-Tipping Hats: 100 Years of Invention in the Pacific Northwest*. Seattle: Sasquatch Books Inc., 1991. Very good.

Index